EXPLORING
THE SOLAR SYSTEM

D1400795

DISCOVERING THE
SOLAR SYSTEM

EXPLORING
THE SOLAR SYSTEM

CAROLE STOTT

BARRON'S

First edition for the United States and Canada published
in 2006 by Barron's Educational Series, Inc.

A QUINTET BOOK

ISBN-13: 978-0-7641-7930-3
ISBN-10: 0-7641-7930-6

Library of Congress Control Number 2005930854

QUIN.GALS

All inquiries should be addressed to:
Barron's Educational Series, Inc.
250 Wireless Boulevard
Hauppauge, NY 11788
www.barronseduc.com

This book was designed and produced by
Quintet Publishing Limited
6 Blundell Street
London N7 9BH

Project Editor Ruth Patrick
Editor Frank Ritter
Layout Jon Wainright
Glossary Illustrations Richard Burgess

Art Director Tristan de Lancey
Managing Editor Jane Laing
Publisher Judith More

Manufactured in Singapore by Pica Digital Pte Ltd.
Printed by Starlite Development (China) Ltd.

10 9 8 7 6 5 4 3 2 1

Contents

Introduction

Humans have looked out from Earth and into space for thousands of years. But they have only truly been exploring space for about fifty. In October 1957, space exploration turned from a dream into a reality with the launch of *Sputnik 1*. It was the first man-made craft to be carried away from Earth and released into orbit around our planet.

Since that time, scores of robotic spacecraft have left Earth behind and moved off into space to investigate it on our behalf. Space probes have journeyed to the planets and their moons, to comets and asteroids, and toward our local star, the Sun. These car- or bus-sized craft, with their armory of instruments and tools, fly by a target, orbit around it, or land on it. Many carry a second, smaller probe, which once released may enter an atmosphere, or land on a surface. Meanwhile, Earth is not forgotten. Science satellites orbiting it today not only look out to space but also down to Earth, monitoring our planetary home.

In this book, we look first at these satellites and the probes sent to our near space neighbors, starting with the Moon and Venus. Then we move on to the missions to the distant giant planets, and to the smaller worlds, the comets and asteroids. Craft from every decade are included, from the first of the late 1950s through to the present years of the twenty-first century, and on to those planned for the future.

My generation has lived through the successes and occasional failures of fifty years of exploration, and now looks forward to new missions and discoveries ahead. For, there are always questions to be answered and challenges to be met. Today, space is an international business involving thousands of people. Together they work on the data collected by past missions, on today's working spacecraft, and on plans for exploration in the decades ahead.

Some will not live to see the completion of their work. It will be the next generation of people who will benefit; people like you. And your vision and curiosity, and that of each generation that follows will continue mankind's exploration of space. Don't simply dream about it, be a part of it.

Carole Stott.

◀ U.S. astronaut Steven L. Smith, tethered
  to a robotic arm, works high above Earth.

CHAPTER 1
Orbiting Earth

Humans have been learning about Earth, the third planet from the Sun, for thousands of years. Many aspects and features of its landscape and environment have been studied and are known in great detail, but large-scale phenomena are neither easy to identify nor easy to study while on the surface. The Earth has to be investigated from space along with the other solar system planets. Spacecraft orbiting Earth can look down from their vantage point some hundreds of miles above the surface and see the planet in its entirety. They can identify large-scale trends as well as zoom in for detail. These Earth-observing satellites offer a new way of seeing our world. The data they collect has already brought a better understanding of our planet and led to better management of its resources.

Earth orbiting satellites

Earth observation satellites that look down on Earth are just some of the many hundreds of satellites that are orbiting around our planet at any one time. Additionally, there are communication satellites that supply television programs and relay phone calls; navigation satellites that help ships and aircraft find their way around the world; and weather satellites that provide data for regional and larger-scale weather forecasts. There is also a small number of satellites looking out from Earth and into space. These minivan-sized scientific satellites are commonly called space telescopes, or space observatories. Since about 1965 astronomers have been using them to collect and record data, in much the same way as their Earth-based equivalents are used.

Earth observation

Designed, built, and used by many nations around the world, Earth observation satellites observe global and continent-wide phenomena and their effects, such as the development, movement, and widespread destruction wrought by hurricanes. The satellites also provide coverage of areas that are inaccessible from the ground, and, because they are observing over large timescales, they can be used to identify gradual changes such as the depletion of the protective ozone layer in Earth's atmosphere. Their sophisticated instruments are

◪ The orbiting x-ray telescope XMM-Newton collects data on objects in deep space.

◪ *ERS-2* is prepared for its launch. It has monitored the Earth since April 1995.

An *ERS-1* image shows the extent of an oil spill (dark area) off the Spanish coast.

unaffected by the weather and operate twenty-four hours a day. By sending out radar pulses and recording the ways in which they bounce back from Earth, they measure land and sea height to great accuracy, also discerning surface roughness, which provides information about the nature and form of the surface. Other onboard instruments are there to help identify different chemical elements in the atmosphere.

JERS-1 used radar to survey Earth's land usage, coasts, and surface geology.

Data from missions are shared. The European Space Agency, for example, not only has access to data collected by its own missions, such as ERS-1, ERS-2, and Envisat, but also processes and stores data from other missions. These include still-operational craft such as the television-sized *Proba*, which was launched from Sriharikota, India, in October 2001, and the U.S. Polar Orbiting Environmental Satellites (POES), as well as the U.S. Landsat series of craft. The data also derives from craft that have finished their working lives; from *IRS-P3* (Indian Remote Sensing Satellite); *JERS-1* (Japanese Earth Resources Satellite); and the U.S. *Nimbus-7*, which for a decade returned daily maps of ozone levels in the atmosphere.

Meteorological data

Some craft are designed to investigate specific issues. *TRIMM* (Tropical Rainfall Measuring Mission), a joint mission between the United States and Japan, has been monitoring tropical and subtropical rainfall through microwave and visible infrared sensors for more than eight years. Since 2002, *Aqua*, which is a joint mission of the United States, Japan, and Brazil, has been collecting data about the Earth's water cycle. This data will, among other things, determine whether the water cycle is accelerating. In 2003, the U.S. *ICESat* (Ice, Cloud, and land Elevation Satellite) started to monitor ice sheets in a bid to help scientists understand the link between changes in the climate and in ice sheets and sea levels. And *Double Star* is studying the effects of the Sun on Earth's environment. *Double Star* consists of two craft that were launched separately by China in 2003 and 2004, but are working together in space. One is in polar orbit around Earth, and the other is close to Earth's equator. Half of the instruments on board the Chinese satellites are European.

Planetary investigation

New satellites are under development for launch in the years ahead. Some will continue the work of those currently in orbit and others will address new issues. The European *SMOS* (Soil Moisture and Ocean Salinity) for instance, once launched in 2007, will contribute to our understanding of the planet's water cycle. And the European Swarm program, which consists of three satellites working together and is scheduled for launch in 2009, will study Earth's magnetic field and improve our knowledge of Earth's interior and climate.

European Remote Sensing

Europe's first Earth observation satellite, *ERS-1* (European Remote Sensing), was launched in 1991. Its core equipment was an infrared imaging sensor and two radar instruments. In 1995, *ERS-2* joined *ERS-1* in space and eventually took over its role. This second craft was identical apart from its additional sensor for monitoring ozone levels in the atmosphere. The two worked together for nine months during 1995 and, because they orbited only twenty-four hours apart, were able to deliver data that showed changes over a particularly short period of time. *ERS-1* stopped working in May 2000; *ERS-2* continues to observe the globe, orbiting around it every 100 minutes. Satellites such as *ERS-2* carry out routine monitoring of Earth's land, oceans, and polar caps. Their

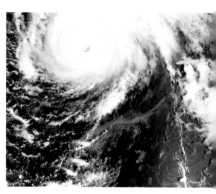

🔲 Hurricane Rita is imaged by the MERIS instrument onboard *Envisat* in 2005.

observations provide valuable information in times of disaster; for example, *ERS-2* was used to identify damage caused by the Indian Ocean tsunami of December 2004.

Envisat

Envisat is the largest Earth observation satellite ever built. This European craft was launched on March 1, 2002, aboard an Ariane 5 rocket from Kourou in French Guiana. It is orbiting Earth every 101 minutes at an altitude of

500 miles (800 km). It repeats its cycle of observations every thirty-five days, although many of its instruments see the entire planet every four days. There are ten optical and radar instruments on board, which collect data on Earth's land, oceans, ice caps, and atmosphere. Every three days the MERIS (Medium Resolution Imaging Spectrometer) instrument completes a new global view of the planet and, for example, its measurements of the color of the sea and coastal areas are used to identify pollution. During 2005 *Envisat* collected data for the most detailed picture of Earth's land surface to date. Once fitted together the data will produce a map of the world of unprecedented resolution. The volume of information generated is overwhelming, the equivalent to that contained in 20 million books.

Landsat

The Landsat series of satellites has been studying Earth's land features for more than three decades and has taken images of the same areas decades apart. There have been seven craft in the series. The most recent, *Landsat 7*, was launched in April 1999. The minibus-sized satellite works about 440 miles (700 km) above the ground. It orbits every ninety-nine minutes and makes a complete cycle of the Earth's surface every sixteen days.

Available information

Although Landsat is U.S. owned and operated, the information it collects is available to both governments and private organizations in countries around the world. Pakistan and Australia have used Landsat data to assess flood damage in their countries. Landsat data has revealed hitherto unknown mountains in remote areas of Antarctica, as well as changes along its coastline. Closer to home, it has identified nearly 9,000 fields growing twenty-five different crops in one California valley. Such information is

Landsat images were used for this view of the Tehachapi Mountains in California.

Rainforest destruction is seen in the lower part of this false-color Landsat image.

used for accurate crop forecasting, which helps to plan labor, fuel, and transportation requirements.

Terra

Terra was launched from Vandenberg Air Force Base, California, on December 18, 1999. By February of the following year it was in its polar orbit and its five sensors were ready to start work. *Terra* is the first in a series of craft that will collect eighteen years' worth of global data. This particular craft will work for at least six years. Canada and Japan have joined the United States in this venture, which is designed to study interactions in Earth's atmosphere, land, and oceans.

Looking into space

One of the main advantages of space-based over ground-based telescopes is that they can work all day and every day. Their use is restricted neither by the daytime Sun nor by the Earth's cloudy atmosphere; they have a clear, uninterrupted view into space. Space-based telescopes also work in a range of wavelengths; some wavelengths, such as x-ray, do not penetrate Earth's atmosphere and so can be collected only from space. By collecting views using x-ray, infrared, ultraviolet, and other wavelengths, astronomers add new dimensions to familiar objects and reveal otherwise invisible phenomena associated with them. They build up a more complete picture of the universe and also learn about its past and how it may evolve in the future.

■ A 2003 Ikonos satellite view of the lush vegetation of Aceh, Sumatra, Indonesia.

■ The same coastline as above, after being swamped by a tsunami in December 2004.

The first space telescopes collected ultraviolet wavelengths and were launched in the 1960s. The first to record x-rays followed in 1970; then came others to work in gamma ray, infrared, and microwave wavelengths. As these early satellites completed their working life they were replaced by newer versions. Today's instruments include the European XMM-Newton telescope, which has been collecting x-rays since December 1999, and the

🖾 The Spitzer telescope views the remnant of a supernova—an exploding star.

Spitzer Space Telescope, which was launched in 2003 and sees the infrared universe. These and others will in turn be replaced by new versions, some of which are already in production and others of which are in the planning stage. In the years ahead two satellites, *COROT* and *Kepler*, will look for rocky planets around distant stars; *Herschel* with its 11.5-foot (3.5-m) mirror, the largest ever to be launched into space, will observe the infrared universe; and

🖾 The space shuttle's robotic arm captures the Hubble Space Telescope from its orbit.

Planck will monitor the cosmic background radiation in order to improve our understanding of the birth and evolution of the universe.

Principally, the work of space telescopes is to look out of the solar system and into deep space, into the realm of the stars and galaxies. Their worth is in uncovering the secrets of these distant objects. But occasionally they are turned on to objects within the solar system. This has been the case with the Hubble Space Telescope, possibly the most widely known of all space telescopes.

Hubble Space Telescope

The space shuttle *Discovery* launched the Hubble Space Telescope on April 24, 1990. The telescope, often commonly called Hubble, orbits about 375 miles (600 km) above Earth's surface, circling around once every ninety-five minutes. Since its launch it has traveled more than 3 billion miles. It was designed in the 1970s and has a primary mirror measuring 8 feet (2.4 m) in diameter. Much of the telescope was built in the United States and parts, such as the solar panels and cameras were built in Europe.

Serviced by space shuttle

The whole instrument was designed for routine servicing by visiting astronauts aboard the space shuttle, and it has undergone four servicing missions during its lifetime. The first, in December 1993, was to correct a problem with the main mirror that became apparent only when the first images were sent back to Earth. Upgrading, maintenance, and replacement of worn-out parts were also carried out in February 1997, December 1999, and March 2002. Another servicing mission scheduled for January 2004 was called off because of safety issues with the shuttle. It is possible that this may still go ahead. Either way, Hubble will simply continue until it is unable to work or is turned off.

Looking back in time

Hubble has observed more than 25,000 astronomical targets during its working life. The vast majority of these have been stellar objects. Notably it has captured detailed views of star-forming nebulae, and of young stars with planet-forming discs, as well as stunning images of planetary nebulae forming as dying stars throw off their outer layers. The telescope has peered back in time to see the first galaxies, and has looked closely at black holes and quasars. Each year more than 1,000 astronomers from around the world submit requests for telescope time; only about 300 are successful. Those who have made observations within the solar system have imaged aurorae around the polar regions of Jupiter and Saturn, recorded the first detailed images of Pluto and its moon Charon, and watched Jupiter's vast storm, the Great Red Spot, change over the years. In 1994 Hubble caught on camera the fragmented comet

Shoemaker-Levy 9 crashing into Jupiter, and on July 4, 2005 Hubble was pointed toward comet Tempel-1 to witness the impact of a projectile from the *Deep Impact* spacecraft.

James Webb Space Telescope

Hubble's replacement, the James Webb Space Telescope, is due for launch in 2013 at the earliest. At 20 feet (6 m) in diameter, its mirror will be larger than Hubble's and will have a greater

🔲 One of the hexagonal segments of the James Webb Space Telescope mirror.

light-gathering power, enabling it to see more. But this is not the telescope's largest feature. A tennis court–sized sunshield will keep the Sun's glare off the mirror. The mirror and shield will be folded for launch and opened for use only once in space. The telescope will work about 1 million miles (1.6 million km) from Earth, collecting information primarily in the infrared but also at visible wavelengths. It will pay particular attention to the earliest galaxies and some of the first stars.

CHAPTER 2

Exploring Close to Home

The Moon is the closest space body to Earth, so it is natural that humans would want to explore it first. It has been a source of fascination for centuries, and humans from diverse walks of life have dreamed of traveling there. Their dreams started to come true a little more than 100 years ago when realistic ideas of space travel were first developed. The scientists and engineers moved from drawing-board theory to the practice of building and launching rockets in the 1920s. Soon, they had a rocket powerful enough to escape Earth's gravity and make it into space. Next came rockets that carried cargo into space, opening the gateway to space probe exploration. In 1957, the Soviet Union launched the first man-made spacecraft into orbit around Earth. *Sputnik 1*, as it was called, circled Earth every ninety-six minutes and signaled "bleep-bleep" for twenty-one days.

The race to the Moon

The space age had dawned, and the two countries with space-travel capability, the United States and the Soviet Union, had ambitious plans centering on the Moon. More than sixty craft have, to date, journeyed to the Moon, most of them sent by those two countries; most flew during the decade from 1959 to 1969. The majority of them were robotic craft; only seven, all of them in the U.S. Apollo series,

carried crew. Many of the early spacecraft were sent to gather the information needed for a successful human landing on the Moon, which the United States achieved in 1969. The Soviet Union abandoned its plans to walk on the lunar surface, although it sent probes that returned with samples of the Moon's soil. Also, two Soviet robotic craft, *Lunokhods 1* and *2*, roved more than 29 miles (47 km) of the lunar surface in the early 1970s.

The Moon revisited

Compared with the frenetic activity seen in the 1960s and early 1970s, the 1980s were quiet in terms of lunar exploration—no craft were sent from Earth at all. Space exploration of the Moon was restarted in the 1990s, when four craft were sent moonward. In the first decade of the twenty-first century, at least six craft are in operation, or are planned for launch. These recent craft have included the U.S. probes *Clementine* and *Lunar Prospector* but

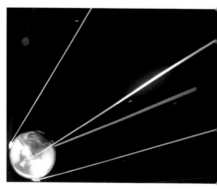

◻ *Sputnik 1*, the first man-made satellite, orbited Earth every ninety-six minutes.

◁ Three remote sensing instruments on *Smart-1* scan the Moon's surface.

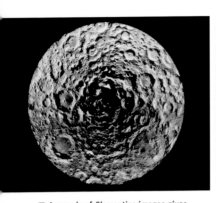

🔲 A mosaic of *Clementine* images gives a view of the Moon's south polar region.

🔲 ESA's *Smart-1* is integrated into an Ariane 5 rocket for its journey to the Moon.

also include the first non-U.S. or non-Soviet craft to the Moon. The first of these, and the first to come from Japan, named *Hiten*, was sent in 1990; the first European craft, *Smart-1*, was launched in 2003; and both China and India now have the Moon in their sights.

The first lunar mission

The first craft ever to escape Earth's gravity and move off into space headed for the Moon. It was the space probe *Luna 1*, which started its journey from Earth on January 2, 1959. Although it was intended to hit the Moon, a control-system failure meant that it flew within 3,500 miles (5,630 km) of its target and moved off into orbit around the Sun. *Luna 1* was the first in a Soviet series of Luna missions that ended with *Luna 24* in 1976. Subsequent Luna craft orbited and landed on the Moon, and returned to Earth with lunar samples.

Soviet exploration

Luna 2 was the first probe to hit the Moon, or, indeed, to land on any celestial body. It crashed into the Moon on September 13, 1959, to the east of Mare Serenitatis. The third probe in the series, *Luna 3*, which followed just two weeks later, was also successful. It kept to its figure-eight path, which took it around to the far side of the Moon. This side never faces Earth and so is unseen. It was sunlit at the time that *Luna 3* flew by, and the probe captured seventeen images of it. By comparison with today's sharp images, or even those taken only a few years later, the *Luna 3* images were indistinct. But they were clear enough to show a dark area that was later named Mare Moscoviense. The Soviets returned much sharper images of the Moon's far side in 1965. Their *Zond 3* probe failed in its aim of flying by Mars but passed the Moon, and as a communications test recorded twenty-five images of the far side.

The next craft in the Luna series used a more sophisticated and, it was hoped, a more accurate strategy to reach the Moon. Rather than launching straight at the Moon from Earth, they were put into a temporary orbit around Earth, from where they fired a rocket to put them on their path to the Moon. The plan was that *Lunas 4* to *8* would all make controlled soft landings. *Luna 4*, launched on April 2, 1963, missed its target by thousands of miles, as did *Luna 6* in June 1965. Just one month earlier, in May 1965, *Luna 5* crashed on the lunar surface. This was also the fate of *Lunas 7* and *8*, in October and December 1965 respectively.

Successful Moon landings

The following year, however, started off successfully. *Luna 9* soft-landed at Oceanus Procellarum on February 3. A spherical lander was released from the mother craft, which had fired its main rocket to slow its speed. This capsule bounced on to the surface, rolled to a stop, and then opened its four petal-like covers and extended its radio antennae. Over the next three days it transmitted data back to Earth from its television camera. These were the first close-up images of another world. The panoramic view showed the rubble-strewn surface of the landing site and the distant horizon.

Intercepted image

The British astronomer Bernard Lovell was one of the first to see the images, and then to make them available to a non-Soviet audience through publication in his local newspaper. He was following the *Luna 9* mission using the U.K.'s Jodrell Bank radio telescope at the Jodrell Observatory in Cheshire. He suspected that a change in the signal he was receiving was an image being sent to Earth. The signal was fed into a borrowed teletype machine, and much to everyone's delight it produced an image of the Moon's surface.

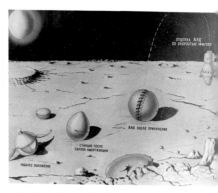

The *Luna 9* lander detaches, lands, separates from its casing, and opens.

Luna 9 on the lunar surface. The lander kept in touch with Earth for three days.

First Moon orbit

Just weeks later, in April 1966, the Soviets achieved another space first when *Luna 10* went into orbit around the Moon. Significantly, it found that the radiation in lunar orbit would not be harmful to any human explorers who might travel there. Further orbiters followed this April success: *Lunas 11* and *12* in August and October 1966, and *Luna 14* in April 1968. *Luna 13*—a lander—arrived in Oceanus Procellarum on December 24, 1966.

Analyzing the soil

Not only did *Luna 13* successfully soft-land on the lunar surface and then transmit images of its surroundings back to Earth, but it also used its two mechanical arms to determine the density and chemical composition of the surface. By 1968 the Soviets had collected enough data with the aid of these probes to make informed selections of landing sites for their manned missions.

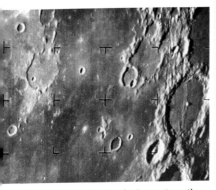

■ Mare Nubium taken by *Ranger 7* was the first U.S. close-up picture of the Moon.

Ranger missions

The first series of craft intended to provide the United States with close-up views of the lunar surface was named *Ranger*. It was also the first U.S. space program to investigate a celestial body. There were nine craft in all, but only the final three in the series were successful. The craft carried television cameras that would operate as a craft approached the Moon, up to the point where it crashed into the lunar surface. Solar panels on either side of a craft's body powered the cameras and provided the power for transmitting the images to Earth. *Rangers 1* and *2*, launched in 1961, were test missions, but both failed when their rocket systems sent them into the wrong orbits. *Rangers 3 to 5* were designed to return data from the Moon and release a capsule that would land on the surface. All three had equipment failures, and the lander idea was dropped for the later missions.

Crash landings

Ranger 6, launched on January 30, 1964, had a good start and crashed into Mare Tranquillitatis as planned, but it failed to send any data from the surface. The Ranger series had its first success six months later, when *Ranger 7* flew in late July 1964. It was launched on 28 July, and in the thirteen minutes of its mission, before it crash-landed into the Mare Nubium on July 31, its six television cameras returned more than 4,000 images. *Ranger 8* sent 7,000 images on February 20,

1965 before crash-landing in Mare Tranquillitatis; and *Ranger 9,* 5,800 images on March 24, 1965, before hitting the floor of Alphonsus Crater. The high-quality images showed surface features less than 3.2 feet (1 m) across.

Preparing for landings

The United States, like the Soviet Union, was sending craft to the Moon to learn more in preparation for achieving their ultimate aim of sending manned missions. Five craft under the series name of Lunar Orbiter were launched between August 1966 and August 1967. Their purpose was to look for safe landing sites in the region around the equator, between 43 degrees E and 56 degrees W. All five were successful, and between them they photographed 99 percent of the lunar surface. The craft imaged the surface for just a few days each. The images were radioed back to Earth, where they were built up strip by strip into a composite whole.

Lunar Orbiter

Lunar Orbiter 1 was the first U.S. craft to orbit the Moon. It and *Lunar Orbiters 2* and *3* followed low-altitude orbits and were focused on imaging possible lunar landing sites that had been earmarked for investigation from Earth. *Lunar Orbiters 4* and *5* flew high-altitude polar orbits and between them imaged the near and far sides of the Moon. All five of the craft were crashed into the lunar surface once their work was finished, the first four

◻ *Lunar Orbiter 2* was flying 28 miles (46 km) above the Moon when it took this image.

so that they could not get in the way of the next probe to arrive on the scene.

Surveyor

The seven U.S. Surveyor probes that traveled to the Moon between June 1966 and January 1968 were all landing craft. The Surveyors were investigating the Moon from the

◻ The Atlas-Centaur 10 rocket carrying *Surveyor 1* lifts off from Cape Canaveral.

surface at the same time that the Lunar Orbiter missions were scrutinizing it from above. Two compartments of instruments were housed within each Surveyor's triangular-shaped framework. A shock-resistant foot supported the craft at each of its three corners, and a solar panel and a communications antenna were attached to a pole rising up from the craft's center. A retro-rocket in the base slowed the craft's descent to the lunar surface. Once it was on the surface, a television camera recorded the view and a collector tool dug into the surface for samples to test.

Mixed fortunes

The first in the series, *Surveyor 1* was a resounding success. It started on its journey on May 30, 1966, and soft-landed on the Moon in Oceanus Procellarum on June 2. It tested the soil and returned more than 11,000 images of the surface, including images of its own landing gear, which showed

that the craft had not sunk into the surface as some had feared. This first success was repeated by *Surveyor 3* when it transmitted 6,000 images from the same region in April 1967 and dug four trenches in its landing site. *Surveyors 2* and *4* did not fare so well. They both crash-landed rather than soft-landed on to the surface.

Experimental liftoff

Surveyor 5 made a soft landing at Mare Tranquillitatis on September 10, 1967. Its soil chemical analyzer found basalt rock—similar to volcanic rock on Earth—and its cameras transmitted more than 19,000 images. This was followed just two months later by the sixth craft in the series. *Surveyor 6* made a soft landing in Sinus Medii on November 10, and achieved a first in space exploration seven days later when it fired its engines and made the first lunar liftoff. The probe rose about 10 feet (3.7 m) above the surface and moved 8 feet (2.4 m) in a westerly direction before touching down. It sent more than 30,000 images. The last of the series, *Surveyor 7*, soft-landed near Tycho Crater on January 10, 1968 and also analyzed soil and returned images.

Ready for landings

The Ranger, Lunar Orbiter, and Surveyor craft all prepared the way for a manned lunar landing by the United States. The Moon had been mapped, potential landing sites had been investigated, and craft had tested the landing conditions. It had been

Surveyor 7 landed just north of the bright ray crater Tycho (lower center).

shown that it was possible to soft-land on the Moon and then to move off its surface, an operation that was essential if the astronauts were to come home safely. The way was thus ready for the Apollo manned missions to the Moon.

Manned Moon landings

The first of the Apollo missions were flown while the last of the Lunar Orbiter and Surveyor probes were completing their work. These tested the equipment that would eventually go to the Moon. But none of them, including the first piloted flight, *Apollo 7*, left the vicinity of Earth. *Apollo 8*, launched on December 21, 1968, was the first manned craft ever to leave Earth's gravity and the first to orbit the Moon. The three astronauts, Frank Borman, James Lovell, and William Anders, completed ten two-hour orbits of the Moon and photographed potential landing sites before returning home. In March 1969, the crew of *Apollo 9* tested the new-style space suits to be worn on the lunar surface, and the lunar module, in this case called *Spider*, that would take the astronauts to the Moon's surface.

The next mission, *Apollo 10*, was a full dress rehearsal for a manned lunar landing. The craft was launched on May 18, 1969 and over the next eight days tested every part of the system. The lunar landing itself was simulated by two of the three-man crew. Thomas Stafford and Eugene Cernan flew the lunar module *Snoopy* to within a few miles of the lunar surface before

Apollo 9's crew. From left: James McDivitt, David Scott, and Russell Schweickart.

David Scott stands in the *Apollo 9* lunar module during its test flight over Earth.

returning to join John Young, who was orbiting in the command module.

Apollo 11

The crew who were to make the first manned landing on the Moon started their journey from Earth at 4:15 A.M. on July 16, 1969. After a brief medical examination and a breakfast of steak, eggs, and toast, the three men started to prepare for their long day ahead. The commander of the *Apollo 11* mission, Neil Armstrong, the lunar module pilot, Edwin "Buzz" Aldrin, and the command module pilot, Michael Collins, arrived at the suiting-up room at 5:30 A.M. An hour later they were on their way to the launch pad where the mighty Saturn V rocket was ready for launch. At 8:32 A.M. they blasted off on their 251,000-mile (403,900-km) journey, watched at the site by one million spectators who had traveled to Cape Canaveral, Florida, for the event.

◪ Buzz Aldrin, second man on the Moon, is photographed by the first, Neil Armstrong.

◪ The Control Room team in Houston, Texas, celebrates *Apollo 11*'s return to Earth.

Separating from *Columbia*

Apollo 11, which consisted of three parts, occupied the nose cone of the rocket. Once above Earth it was these three parts, locked together, that headed for the Moon: the command module, called *Columbia*; the service module; and the lunar module, named *Eagle*, which would carry two of the crew to the lunar surface. Once in orbit around the Moon, *Eagle* separated from the rest of the craft and moved on to a course that took it closer and closer to the lunar surface. Inside, Armstrong and Aldrin stood at their flight stations. The descent engines slowed their approach and Armstrong took over the controls to make a safe touchdown, even though he found it difficult to see because of the clouds of dust blown up by the craft.

Landing of the *Eagle*

The *Eagle* landed in Mare Tranquillitatis on Sunday, July 20, 1969; the time at mission control in Houston, Texas, was 3:17 P.M. Once they had donned their outdoor suits and depressurized the cabin, they opened the hatch and Armstrong climbed down the *Eagle*'s ladder. His next words are now famous: "That's one small step for man—one giant leap for mankind."

Nineteen minutes later Aldrin stepped on to the surface. The two moved around using kangaroo-hops to compensate for the Moon's gravity, which is one-sixth that of Earth. Over the next two hours or so they scooped up dust and took soil samples from below the surface; they planted the U.S. flag, took photographs, and collected samples of the solar wind. They set up two pieces of scientific equipment: a seismometer to detect moonquakes, and a retro-reflector for use in determining the distance from the Moon to Earth, by timing how long it takes for a laser beam to be reflected back to Earth. On Earth one-fifth of the population, about 600 million people, watched the two men. The television images were relayed by a camera that Armstrong had set in position on board *Eagle* before moving down the ladder, and another that they had set up on the Moon's surface.

The *Apollo 11* crew leave *Columbia* after the splashdown in the Pacific Ocean.

Returning to Earth

Just twenty-one hours and thirty-six minutes after arriving, Armstrong and Aldrin blasted off from the Moon in the upper part of the lunar module. It bore them away from the surface to be reunited with *Columbia* and Collins, who meanwhile had been orbiting the Moon once every two hours. Within four hours the *Eagle* had docked. The astronauts and their cargo, including the 48.4 lb. (22 kg) of Moon rock they had collected, were then transferred between the craft before the *Eagle*'s upper stage was released to fall back on to the surface. Once the engines had placed the remaining Apollo craft— the combined command and service modules—on to their path to Earth, the crew took a well-deserved rest.

When they neared their home planet, the service module was jettisoned. *Columbia* and the crew then headed through Earth's atmosphere toward the Pacific Ocean. Parachutes slowed the craft for its splashdown, and the men were soon aboard the recovery ship USS *Hornet* and heading for Hawaii. From there they flew to Houston, where they spent three weeks in an isolation unit to reduce the possibility of contamination from any microorganisms that they had unknowingly brought from the Moon. They were welcomed as global heroes.

The United States had achieved their aim, set out by President John F. Kennedy on May 25, 1961, of "landing a man on the Moon and returning him safely to Earth" before the end of the 1960s. Their success came after the many space firsts achieved by the Soviet Union—the first spacecraft into Earth orbit, the first man in space, the first woman in space, the first space walk—and the Americans were thrilled and triumphant to be first to walk on the Moon.

Moonwalks

Over the next three and a half years, eighteen more astronauts were involved in six more Apollo missions to the Moon. These concentrated on making a scientific investigation of the lunar landscape. *Apollo 12* arrived at the Moon just months after *Apollo 11*. Charles Conrad and Alan Bean landed in Oceanus Procellarum as Richard Gordon orbited overhead. Their first steps on the Moon were watched from

■ John Swigert works to reduce the high carbon dioxide level inside *Apollo 13*.

■ Alan Shepard and the "rickshaw" that carried his rock-collecting equipment.

Earth, but sunlight damaged the television camera's lens, and no images could be taken for the rest of the time on the lunar surface. *Apollo 12*'s lunar module *Intrepid* had landed close to the site of the *Surveyor 3* landing craft, and the second of Conrad and Bean's two moonwalks included a hike to *Surveyor 3*. They retrieved its camera and other parts for return to Earth, where they were examined for effects of their time on the Moon. They also brought back 75 lb. (34 kg) of lunar rock and soil.

Unlucky *Apollo 13*

An oxygen tank in the service module exploded as *Apollo 13* was on its way to the Moon in April 1970. The craft had to continue toward the Moon even though the plan to land was aborted. The priority now was to return the three astronauts, James Lovell, Fred Haise, and John Swigert, safely back to Earth. An ingenious recovery plan was worked out and, after it circled the Moon once, the craft was put on a course that brought the crew home.

In light of the *Apollo 13* mission, the *Apollo 14* flight was delayed some weeks while changes were made to the craft and flight plan. *Apollo 14* landed on February 5, 1971 in the Fra Mauro region, the intended landing site of *Apollo 13*. Alan Shepard and Edgar Mitchell completed two walks when they collected samples and set up equipment. They had a wheeled cart, nicknamed the "rickshaw," to carry their tools and rocks. As they explored, the Soviet robotic craft *Lunokhod 1* was moving elsewhere on the Moon.

Lunar Roving Vehicles

The next three missions, *Apollos 15, 16,* and *17* took a Lunar Roving Vehicle to the Moon. Usually just called the "rover" or "buggy," this vehicle, which resembled a golf cart, enabled the astronauts to cover a greater area of the lunar surface. They ventured up to about 6 miles (9.6 km) from their landing sites. Any farther and they would not have been able to get back to their lunar module if the rover had broken down. The lightweight car was built mainly from aluminum, was powered by batteries, and reached speeds of about 11 mph (18 km/h). An onboard antenna kept the astronauts in communication, and a television camera enabled the Earth-based spectators to follow their progress.

Apollo 15 landed in a relatively flat region of Hadley Rille, near the base of the Apennine mountain range. *Apollo 16,*

⬛ *Apollo 17*'s crew: Eugene Cernan (seated), Harrison Schmitt (left), and Ronald Evans.

by contrast, landed in the Descartes highlands and made the highest manned lunar landing. Both missions included three walks apiece lasting about five to seven hours each. The first scientist on the Moon traveled on *Apollo 17,* the last manned mission. Geologist Harrison Schmitt made three excursions, each more than seven hours long, with Eugene Cernan, who was the last man to walk on the Moon. Their lunar module, *Challenger,* landed

Men who landed on the Moon

Mission	Landed	Astronauts	Time on Moon
Apollo 11	July 20, 1969	Neil Armstrong, Edwin "Buzz" Aldrin	21h 36m
Apollo 12	November 19, 1969	Charles (Pete) Conrad, Alan Bean	31h 31m
Apollo 14	February 5, 1971	Alan Shepard, Edgar Mitchell	33h 31m
Apollo 15	July 30, 1971	David Scott, James Irwin	66h 55m
Apollo 16	April 20, 1972	John Young, Charles Duke	71h 2m
Apollo 17	December 11, 1972	Eugene Cernan, Harrison Schmitt	75h 0m

in the Taurus-Littrow Valley, chosen because it was the site of a landslide. They collected 242 lb. (110 kg) of rock and soil, more than any other crew, and had the longest stay on the Moon, as well as the longest time exploring its surface.

Comparing Moon regions

The Apollo astronauts explored six regions of the Moon and collected samples from its different types of landscape. Rocks and soil were taken from the highlands, from impact craters, and from maria (singular mare), huge craters filled with volcanic lava. Studies of these samples and the data collected by the more than fifty scientific instruments set up on the Moon by all crews have yielded valuable information. They have helped in our understanding of the origin and evolution of the Moon, and continue to be of use in research. The rock and soil samples remain available for testing out new ideas as they arise.

⌂ *Zond 6* bounces off Earth's atmosphere to reduce g-forces felt by future astronauts.

Soviet exploration

Like the United States, the Soviet Union in the 1960s worked toward putting a man on the Moon. The missions of *Zonds* 4 to 8 were an early stage in their bid to fly humans around the Moon—a necessary precursor to a human landing. All five craft were unmanned, but *Zond 5*, which was the first-ever flight to journey around the Moon and come back to Earth, did carry passengers. On board, in September 1968, were turtles, flies, worms, plants, seeds, and bacteria to test any potential hazards of space flight. *Zonds* 6, 7, and 8, which launched between November 1968 and October 1970, all flew around the Moon and returned images of its surface. But within six weeks of the first of these missions the United States won the race to send men around the Moon, and only six months later they landed the first humans.

Soviet lunar landings

The Soviet Union then abandoned any plans to send men and put all their efforts into exploring the Moon with robotic craft. Just a week before the first Americans walked on its surface, the Soviets launched the probe *Luna 15*, which was to make a soft landing and return to Earth with soil samples. Its descent, which was too fast and marked the end of this particular mission, took place while Armstrong and Aldrin were on the Moon. By contrast, *Luna 16* was a great success. It landed at Mare Fecunditatis on September 20,

1970 and used its extendable arm to drill 14 inches (35 cm) into the surface and retrieve 3.5 oz. (100 g) of rock and soil. The upper part of the probe launched itself with the samples in a sealed container and returned them to the Soviet Union on September 24. Four more missions were sent to collect and return samples. *Luna 18* and *Luna 23* failed, but *Luna 20* and *Luna 24* were successful. *Luna 20* collected 1 oz. (30 g) of sample from the Apollonius highlands in February 1972, and *Luna 24*, the last of the Luna series of missions, returned with 6 oz. (170 g) from Mare Crisium in August 1976.

🖼 *Lunokhod 2*, an improved version of *Lunokhod 1*, worked by Mare Serenitatis.

Soviet rovers

The Soviets scored a space first with their *Luna 17* mission. On board was *Lunokhod 1*, which was to become the first rover to operate on another world. It landed in the Mare Imbrium, rolled down the ramp of its mother craft, and then started work on November 17, 1970.

The 24-foot (7.2-m) long eight-wheeled rover was a mobile laboratory, which moved about the lunar surface and worked at different sites. Over a ten-month period its Earth-based controllers directed it over 6.5 miles (10.45 km). On board were cameras, soil-analysis tools and other equipment, solar cells to produce power, and communications gear. The rover tested the soil at more than 500 sites and took more than 20,000 images. A second rover, *Lunokhod 2*, arrived aboard *Luna 21* on January 15, 1973,

on this occasion in Mare Serenitatis. It covered more ground, 23 miles (37 km), but worked for a shorter time, less than five months, and returned more than 80,000 images. Meanwhile, two orbiters were following paths around the Moon. *Luna 19* started its study of the Moon's gravitational and magnetic fields in October 1971. Then, in June 1974, *Luna 22* took television pictures of the surface as it orbited overhead.

Japan flies to the Moon

Luna 24's return to Earth marked the end of the busiest period of space probe exploration on any celestial body. It would be more than thirteen years before spacecraft were once again sent to the Moon. On January 24, 1990, Japan became the third nation to send a spacecraft to the Moon with the launch of its flyby probe, *Hiten*. On board was a second craft named *Hagomoro*. Once released it entered lunar orbit, but its transmitter failed to work, and no data was returned.

Clementine

The U.S. craft *Clementine* had two targets in its sights. It was launched moonward on January 25, 1994, where it was to make scientific observations of the Moon as well as test spacecraft components. Once this part of the mission was over, it was to head toward the asteroid Geographos to determine its size, shape, and other properties. The first part of the mission went as planned, and *Clementine* completed its two-month mapping program when it took images of the lunar surface from 60 degrees north to 60 degrees south. It used five different imaging systems, which worked in ultraviolet, optical, and infrared wavelengths with a series of filters, and returned a wealth of lunar images. These included the first color view of the Moon on a global scale. The data collected was used to map the distribution of rock and soil types on the lunar surface. *Clementine* left lunar orbit on May 7 in preparation for the second part of its mission. Unfortunately, one of the onboard computers caused a rocket thruster to fire until it had run out of fuel. The craft was left spinning eighty times per minute, and the flyby of Geographos planned for late August 1994 had to be abandoned.

Lunar Prospector

The *Lunar Prospector* mission lasted eighteen months, from the craft's launch on January 7, 1998, to its demise in the Moon's south polar region. Key elements of the mission were to study the surface composition of the Moon, and to search for deposits of ice at the south pole. The drum-shaped craft was covered in solar cells to produce power. It carried six instruments and an antenna for communication; there was no onboard computer, so the craft was controlled totally from the ground. One of the spectrometers on board recorded concentrations of hydrogen at both the north and south poles, which some have taken as an indication of water ice in permanently shadowed craters. The mission ended on July 31, 1999, when the probe was deliberately sent on a crash course to a permanently shadowed area of an impact crater near the south pole. It was suspected that the crater might contain ice. Earth-based telescopes were pointed to the region to observe the expected cloud of water vapor released from the

◩ Earth above the Moon: just one of the 1.7 million images returned by *Clementine*.

ice deposits as *Lunar Prospector* crashed, but none was seen.

Europe goes to the Moon

Smart-1 is the first, and to date only, European spacecraft to go to the Moon. It also has the role of testing new technology, in this case solar-electric propulsion for use on bigger and deep-space missions. Sweden and five other European countries were involved in building the craft, and scientists from nine European countries were involved in both the design and the use of the instruments on board. European Space Agency science and technology staff based in The Netherlands oversaw the whole project.

Answering new questions

The box-shaped probe was launched on an Ariane 5 rocket from Kourou, French Guiana, on September 27, 2003, along with two large satellites. It is about 3.3 feet (1 m) cubed with 46-foot (14-m) long solar panels extending from opposite sides of the craft. On board are three antennae for communication, a camera, three miniature spectrometers for chemical analysis, and other instruments for monitoring the new propulsion system. Its investigation of the Moon should shed light on the theory that the Moon formed after the collision of a Mars-sized body with Earth, early in the solar system's history. It is also to make an inventory of chemical elements in the lunar surface, and images of the surface taken from different angles will

■ *Lunar Prospector* photographed at an early stage of preparations for its launch.

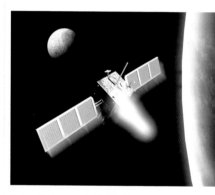

■ *Smart-1* is propelled to the Moon by its innovatory ion drive engine.

be coupled with x-ray and infrared data to produce three-dimensional views. *Smart-1* will also look at crater rims that may always be lit by the Sun, as well as south pole impact craters that never receive the Sun's light. The temperature here is always below −274 degrees F (−170 degrees C) and these craters may harbor water ice. The scientific investigations began in February 2005 and are expected to continue until at least August 2006.

CHAPTER 3

Revealing Venus

Because Venus is the closest planet to Earth, it is an obvious target for space exploration. But it is an incredibly hostile world; it is scorching hot and has a thick blanket of poisonous gas surrounding it. Any craft that reaches the surface has to be able to withstand the temperature of 867 degrees F (464 degrees C) and the high surface pressure, which is more than ninety times greater than Earth's. Like any human observer, a craft orbiting Venus has its view of the surface permanently barred by the blanketing atmosphere of carbon dioxide, with its clouds of sulphuric acid.

Looking beneath the clouds

Although no human can see below the clouds, successful spacecraft missions to Venus have revealed the landscape of this fascinating rock world to us. We have learned something of the composition and structure of its 50-mile (80-km) thick atmosphere, which circulates around the planet every four days. And we know that underneath there is a gloomy, dry world whose landscape has been shaped by intense volcanic activity; much of Venus is covered in volcanic plain, and there are volcanoes all around the planet.

Mariner missions

The first probe to make a successful mission to Venus was part of the U.S. Mariner series. *Mariner 2* was also the first successful probe to a planet. The near-identical craft *Mariner 1* should have flown by Venus first, but its launch was unsuccessful. *Mariner 2* was the first of three Mariner probes to Venus. It flew by the planet on December 14, 1962 after a three-and-a-half-month journey. It found that Venus has a hot surface and cooler clouds. During its journey it also transmitted information back to Earth on interplanetary dust and the solar wind.

Mariner 5

Five years passed before a second probe in the Mariner series was sent to Venus. *Mariner 5* was planned as a backup to *Mariner 4*, and was originally destined for Mars. When *Mariner 4* was a success, *Mariner 5* was prepared for a trip to Venus. It flew within about 2,500 miles (4,000 km) of the planet on October 19, 1967, confirmed the findings of *Mariner 2*, and recorded additional information about Venus's atmosphere and magnetic field.

An impression of *Venus Express*, the latest of over twenty probes to orbit the planet.

Staff present a model of *Mariner 2* to President John F. Kennedy (right).

🔲 **The Mariner craft confirmed that Venus's thick clouds are mainly carbon dioxide.**

Mariner 10

The third Mariner probe to Venus, *Mariner 10*, was programmed to investigate two planets—the first mission to do so. Its ultimate target was Mercury, but it was to fly by Venus on the way. It did this on February 5, 1974, and took the first clear, close-up images of the planet's upper atmosphere, and for the first time established atmospheric circulation patterns. The craft then used gravity assist—a means of using the gravity of a planet to propel a craft on to another destination—to continue from Venus to Mercury. This method was used for the first time on this mission but has been used with great effect many times since.

The Venera probes

The most successful series of probes to investigate Venus was the Soviet Union's Venera missions, which included flybys, atmospheric (descent) probes, and landers. The sixteen craft in the series were launched over a period of twenty years; they made the first soft landing on the planet and provided the first images from the Venusian surface. However, not all were equally successful.

Lost contact

Communications broke down with the first two missions, which were designed to fly by the planet. The first of these was launched in February 1961 and the second in November 1965. A handful of other flybys of Venus were attempted by the Soviets in the intervening years, but none achieved their aim. Communications were also a problem for the *Venera 3* mission, the first of the probes to attempt to land on Venus. *Venera 3* crashed into the planet on March 1, 1966—the first craft to reach the surface—but contact was lost as the craft descended.

The *Venera 4* mission marked the start of the scientific investigation of Venus. As its descent probe moved through the atmosphere, in October 1967, it became the first probe to send back information. It was also the first probe to make *in situ* measurements of the atmosphere, revealing its composition of 97 percent carbon dioxide and 3 percent nitrogen. On May 16, 1969, a year and a half later, *Venera 5* released a descent probe into the atmosphere, and then a day later *Venera 6* did the same. These both transmitted information on the atmospheric temperature and pressure for about fifty minutes as they descended. The first lander—*Venera 7*—transmitted during its descent and for twenty-three minutes from the surface. These were the first signals received on Earth from another planet.

Studying the surface

The first study of the surface came with *Venera 8*. Its lander, which slowed its descent by aerobraking and then by parachute, confirmed *Venera 7*'s temperature and pressure readings and revealed that there was enough light at the surface to take images; daytime Venus was like a dull, overcast day on Earth. *Veneras 9*, *10*, *13*, and *14* all landed safely on the surface in the Beta Regio area. The first pair, *Veneras 9* and *10*, arrived within days of each other in October 1965; then *Veneras 13* and *14* landed in March 1982.

Surface images

Each mission consisted of an orbiter and a lander that separated on arrival at Venus. The landers transmitted data to Earth via the orbiters, including the first-ever pictures from the Venusian surface, taken first by *Venera 9*, then by *Venera 10*. *Veneras 13* and *14* were the first craft to analyze the surface; a drilling arm collected a sample, which the spectrometer on board analyzed as being similar to basalts found on Earth. The 360-degree panoramic views taken by the probes, which were the first color images of the surface, showed rocks like those found in basaltic lava flows on Earth.

Veneras 11 and *12* were not so successful; the flyby craft released soft-landers, but the camera covers did not come off as planned after landing and the soil analysis experiments were damaged. They did, however, record what seemed to be lightning flashes in the Venusian clouds. The final craft, *Veneras 15* and *16*, were orbiters; they worked together in 1983. They were both launched in early June that year and arrived within days of each other

◼ *Venera 4* returned information as it descended through Venus's atmosphere.

◼ Soviet technicians work on *Venera 8* inside the assembly and testing workshop.

⬛ An impression of the *Pioneer Venus Multiprobe* releasing its four miniprobes.

⬛ The *Pioneer Venus Orbiter* identified sulphuric acid in Venus' upper cloud layers.

in October. Over an eight-month period they used radar to map Venus from its north pole down to about 30 degrees north. These were not the last Soviet craft to Venus. *Vega 1* and *Vega 2* flew by Venus in 1984 on their way to rendezvous with Halley's comet. They released two balloons and Venera-type landing craft, which recorded temperature and pressure, and monitored the motions of the Venusian atmosphere.

Pioneer Venus Orbiter

The Pioneer Venus mission to the planet was the final flight of the U.S. Pioneer series of craft designed for interplanetary exploration. It consisted of two spacecraft that were launched separately, three months apart. The first to go was the *Orbiter*; it was launched aboard an Atlas-Centaur rocket from Kennedy Space Center in Florida on May 20, 1978. It moved into orbit around Venus on December 4 of that year. The *Orbiter* had twelve instruments on board; a radar mapper collected data on the planet's surface, and others studied the characteristics of the upper atmosphere and investigated the solar wind in the planet's environment. Its orbit took it within 95 miles (150 km) of the surface and as far as 42,000 miles (66,900 km) away; in this way it could obtain both close-up and distant views. It created a global map of the planet and revealed a landscape of volcanic areas, rolling plains, and continent-sized regions. The craft was designed to work for one Venusian year—about eight months—but most of it was still working thirteen years later, when, on October 8, 1992, it moved into the atmosphere to burn up.

Pioneer Venus Multiprobe

The second craft, the *Multiprobe*, was launched three months after the *Orbiter*, on August 8, 1978, and arrived on 9 December that year. During November the probe had split into five separate craft: the probe transporter, commonly called the *Bus*; a large atmospheric entry probe with seven instruments, named the *Sounder*; and three small identical probes carrying three experiments each. The probes entered the atmosphere, and during the hour or so it took to descend toward the surface they returned data on atmospheric temperature, pressure, and composition. They revealed that the atmosphere had a three-layered cloud structure and confirmed the scientific results of previous missions, but they also noted that temperature differences from day to night and from equator to pole were very small.

Magellan

The United States returned to Venus a little more than ten years after *Pioneer Venus* had finished its work. Their *Magellan* mission was an ambitious but successful one. The probe's task was to orbit around Venus and use radar to see through its unbroken atmosphere and map the planet's surface in more detail. It would build on the work of the *Pioneer Venus Orbiter* and *Veneras 15* and *16*, which had used radar to provide the first real look at the planet's surface terrain. It was *Magellan*'s job to make the first comprehensive survey of the planet and with a hugely improved resolution.

Magellan measured about 15 feet (4.5 m) long and carried a 12-foot (3.7-m) high-gain antenna with the dual purpose of radar imaging and communication with Earth. The craft's main structure and its antenna were both spares from the Voyager missions to the outer planets, and other parts were left over after the Galileo mission to Jupiter. Two square solar panels powered the craft's work at Venus.

Magellan started its journey in the cargo bay of the U.S. space shuttle *Atlantis* on May 4, 1989—the first spacecraft to be launched by a space shuttle. *Atlantis* carried it to a low Earth orbit where, once released, it fired a solid-fuel motor that sent it on a path

◫ The *Magellan* probe is released from the cargo bay of the space shuttle.

◫ A *Magellan* image of Sacajawea Patera, one of Venus's hundreds of volcanoes.

🔼 Venusian volcanoes seen by *Venera* (left)
and in greater detail by *Magellan* (right).

around the Sun one and a half times
before arriving at Venus on August 10,
1990. *Magellan* then fired its motor
and moved into polar orbit to begin
mapping the planet.

Mapping Venus

The craft completed one orbit every
three hours and fifteen minutes, flying
over Venus's poles from north to south,
or vice versa. Its orbit was elliptical,
taking it to within 125 miles (200 km)
of the planet and more than 5,000 miles
(8,000 km) away. After recording an
orbit of data it would use the same
antenna to transmit the data to Earth.
In this way the planetary surface was
revealed strip by strip. Venus has the
slowest spin of all the planets; it turns
once in 243 days. Venus turned slowly
under the space probe, and after one
spin of the planet the mapping
sequence was complete.

Two more eight-month mapping
cycles followed, producing detailed
maps of 98 percent of the planet's
surface. The look angle of the radar
differed from cycle to cycle, so slightly
different views of the same location
were combined to produce three-
dimensional views. The images show
light and dark areas, but these are
different from such areas in a snapshot
of Earth's landscape, where they
indicate the reflectivity of light. In
Magellan radar images they indicate
degrees of roughness, where the bright
areas are rougher than the dark areas.
Magellan's images revealed a volcanic
landscape dominated by lava. Impact
craters seen in the surface indicated
that Venus is geologically young; its
surface is about 500 million years old.

Gravity assessment

The probe gathered data on the
planet's gravity during a fourth eight-
month orbital cycle, which lasted
from September 1992 to May 1993.
By monitoring a constant radio signal,
transmitted by *Magellan* as it passed
over different areas of Venus, Earth-
based scientists could build up a gravity
map of the planet. Even then *Magellan*'s
working life was not over. The craft was
used to test a then untried aerobraking
technique whereby a planet's atmosphere
is used to slow or steer a spacecraft.

Magellan's last days

In September 1994, *Magellan*'s solar
panels were extended like the sails of
a windmill and, as the probe moved into
Venus's outer atmosphere, the scientists
measured the torque control required
to keep the spacecraft from spinning.

This experiment had a dual purpose; scientists collected data on the behavior of molecules in Venus's upper atmosphere, and engineers gathered information useful for spacecraft design. *Magellan*'s orbit was lowered for one final time on October 11, 1994. The craft moved into the atmosphere and toward the surface, gathering valuable data as it moved. Contact was lost the next day. The craft is believed to have burned up in the atmosphere.

Venus Express

The European Space Agency (ESA) launched its first probe to Venus on November 9, 2005. *Venus Express* was borne on the back of the great success of ESA's *Mars Express* probe to Mars. Scientists and engineers were eager to reuse the craft's design and some of its spare instruments, as well as some from ESA's cometary mission, Rosetta. Together, the instruments could make a unique study of Venus's atmosphere.

Investigating the atmosphere

Venus Express was carried into space and set on its journey to Venus by a Soyuz-Fregat rocket from the Baikonur Cosmodrome in Kazakhstan. On arrival at the planet, *Venus Express* is to maneuver into its operational orbit. This will take it to within 155 miles (250 km) of the planet and 41,000 miles (66,000 km) when farthest away. Over the following 500 Earth days, the equivalent of just over two Venusian spins, the probe will complete its mapping mission and perform a

global investigation of the Venusian atmosphere. The data collected by the onboard instruments will reveal more about the atmosphere's composition and how it interacts with the planet's surface. It will also contribute to a fuller understanding of the global characteristics of the atmosphere and how it circulates. At present, it is baffling that the planet spins so slowly yet the upper atmosphere takes just four Earth days to whiz around it.

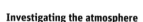 *Venus Express*'s Soyuz-Fregat rocket is made ready for launch at Baikonur.

An impression of *Venus Express* (right) separating from part of its launch rocket.

CHAPTER 4

Toward the Sun

The Sun is by far the closest star to Earth. It is the only one that we can study in detail from our home base, and the only one close enough for investigation by our space probes. But even then we do not want to get too close to the Sun's overpowering energy. By simply placing instruments above Earth we can make better observations of the Sun than from the surface, as well as additional ones that are not possible from the ground. More than a dozen Sun-studying spacecraft have been placed in orbit around Earth. These craft do not stay close to Earth but follow elliptical paths that take them millions of miles away, toward the Sun. A smaller number of craft, however, have moved off into space and observed the Sun from independent solar orbits.

Mariner 10 to Mercury

Mercury is the closest planet to the Sun and because of this proximity is a difficult planet to study from Earth. It is never seen in a fully dark sky and is only ever seen near the horizon, where our planet's atmosphere is particularly turbulent. It is impossible to observe Mercury's surface features from Earth, so scientists at home eagerly awaited the results of the first space mission to this small planet.

Close-up views of Mercury came in the early 1970s when the U.S. probe *Mariner 10* encountered the planet.

The probe revealed a dry, gray-looking world reminiscent of the Moon. Until very recently *Mariner 10* was the only probe to be sent to Mercury. A second probe, also launched by the United States, blasted off for the planet in August 2004. It will, however, be some time before any new data is relayed home. Meanwhile, the European Space Agency has Mercury in its sights; it is preparing the *BepiColombo* space probe, which is planned to set out on its mission to Mercury in 2012.

First visitor

Mariner 10 arrived at Mercury in March 1974 after flying past Venus. It was the first craft to visit two planets, and the first to use the gravitational pull of one to reach another. The maneuver was designed to save time and fuel, and more instruments could be carried on the craft as a result of the reduced fuel load. The probe's eight-sided body had two electricity-generating solar wings projecting from opposite sides.

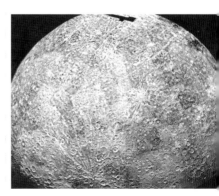

◻ The *Ulysses* craft approaches the Sun. It is the first craft to orbit over the Sun's poles.

◻ Mercury's southern hemisphere is seen in this mosaic of images taken by *Mariner 10*.

■ *Mariner 10* revealed many impact craters, in a range of sizes, on Mercury's surface.

On board were two identical television cameras for imaging the planetary surface, and instruments to measure the physical characteristics of Venus and Mercury. Its high-gain antenna dish was kept pointing permanently at Earth for communications.

Three flybys

Mariner 10 encountered Mercury three times. On March 29, 1974, it flew to about 440 miles (700 km) above the dark side of the planet but took images of the sunlit side of Mercury as it approached and then departed from the planet. On the second flyby, on September 21, 1974, it passed over the south pole at an altitude of 30,000 miles (48,000 km). The last encounter, on March 16, 1975, again came close, just 200 miles (320 km) above the surface, over the nighttime side. The mission ended just eight days later when the nitrogen used to power thrusters for changing the craft's orientation was exhausted.

When *Mariner 10* arrived at Mercury, little was known about this world except some basic properties such as size and the time it takes the planet to spin and to orbit the Sun. The details of its surface terrain were unknown. By the time *Mariner 10* finished work, about 40 percent of the planet's surface was imaged. Everywhere the cameras turned, the surface was riddled with impact craters, and the surface had remained like this for about three billion years. *Mariner 10* also confirmed that Mercury has no significant atmosphere and has a small magnetic field.

Messenger

Mercury was one of the first planets to be explored, but the initial success achieved with *Mariner 10* was not followed up. There was a gap of more than thirty years between the launch of *Mariner 10* and its successor. The second probe, named *Messenger*, was launched from Florida on August 3, 2004 by a Delta rocket. Fifty-seven minutes after liftoff the box-shaped craft was in solar orbit and ready to move its two solar wings into position. Six scientific instruments—an imaging system, four spectrometers, and an altimeter—are mounted on the outside of the body, and a magnetometer is at the end of a long boom. The imaging system will map nearly the entire planet in color, revealing the surface unseen by *Mariner 10*. A curved shield shades the body and instruments from the glare and heat of the Sun.

Messenger spent the first year of its journey orbiting around the Sun. A gravity-assist from Earth in August 2005 set it on a course to pass Venus in October 2006 and again in June 2007. Venus's gravity will change the shape, size, and tilt of the craft's path and bring it closer to Mercury's orbit. Three close flybys of Mercury will follow in 2008 and 2009, when *Messenger* will be just 125 miles (200 km) above the planet. Each of the flybys is followed by a course correction so that in March 2011 *Messenger* will move into orbit around Mercury. Data collected during the flybys will be used to help plan the scientific work that *Messenger* will carry out while it orbits the planet.

Elliptical orbit

The orbital phase of the mission is planned to be one Earth year long, which is approximately equal to two Mercury solar days, measuring from sunrise to sunrise on the planet. The first solar day (equal to 176 Earth days) is scheduled for obtaining global mapping data; the second will be used for targeted science investigations. *Messenger* will follow a highly elliptical orbit reaching from about 125 miles (200 km) above the surface to more than 9,000 miles (15,000 km) distant.

Caloris Basin

The low point on the orbit is over the northern hemisphere and will allow *Messenger* to investigate Mercury's largest known surface feature, the Caloris Basin. This enormous crater was formed about 3.6 billion years ago when a vast space rock collided with the young planet. *Mariner 10* took images of only part of this impact crater, which is 800 miles (1,300 km) wide. In addition to the images, information gathered by *Messenger* will contribute to our knowledge of the size and constitution of Mercury's core and the chemical composition of the surface. The data will throw light on the planet's geological history.

◫ Each space mission has its own patch. Here, *Messenger* flies above Mercury.

◫ The Delta rocket carrying *Messenger* launches from the Kennedy Space Center.

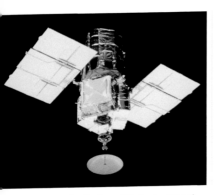

The Solar Maximum Mission is released back into space following its repairs.

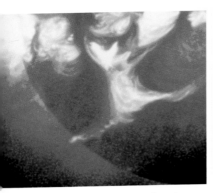

A filament of hot gas erupts from the Sun in this x-ray image by *Yohkoh*.

Studying the Sun

The Sun, and its effect on the space around it and on Earth, has been studied by more than twenty craft. The first mission was in 1962, but most were launched in the 1990s. The missions came from a variety of countries. The United States, Russia, Japan, Germany, and other European countries instigated them, sometimes as a single-nation operation, sometimes as a partnership or multinational venture. The first series of craft designed for observing the Sun was launched by the United States, between March 1962 and June 1975. The eight Orbiting Solar Observatories (OSO) monitored the Sun in ultraviolet, x-ray, and gamma ray wavelengths. They worked together from Earth orbit, one after another, to observe the Sun as it went through a complete eleven-year activity cycle.

Helios and *SMM*

Two German-U.S. craft, named *Helios*, successfully investigated the solar wind in the 1970s. They traveled to within 28 million miles (45 million kilometers) of the Sun, inside Mercury's orbit. In February 1980 the United States launched the Solar Maximum Mission (*SMM*) aboard a Delta rocket from Cape Canaveral into Earth orbit. It was launched at the peak of the solar activity cycle and achieved its aim of studying solar flares, observing more than 12,000 flares shooting up from the Sun's visible surface. In 1984 the crew of the space shuttle *Challenger* retrieved *SMM* and carried out repairs before releasing it back into space. The repair extended the life of the mission by five years; *SMM* worked until November 24, 1989, before burning up in Earth's atmosphere a week later.

Yohkoh and *Geotail*

The 1990s was the busiest and most productive decade of solar exploration. Japan launched *Yohkoh* in August 1991 from the Kagoshima Space Center, to make x-ray and gamma ray observations of the solar corona and solar flares. In its first ten years of operation it took more than six million x-rays of the Sun, which are helping scientists understand how the Sun functions. In 1992 Japan joined with the United States in launching *Geotail* to study solar wind interactions with Earth.

Three U.S. missions—*Wind*, launched in 1994, *Polar* in 1996, and *ACE* in 1997—made similar solar wind studies. Then, in April 1998, the United States launched *TRACE* to observe the solar atmosphere. In the mid-1990s *Interbol 1* and *Interbol 2* were launched by Russia to study the solar wind. These were two pairs of craft, each pair consisting of a Russian craft and a Czechoslovakian craft, with Swedish, French, and Canadian instruments on board. All of these craft of the 1990s worked from Earth orbit.

Cluster

The European Space Agency's first mission to study the solar wind was named *Cluster*. Unfortunately it was destroyed when the Ariane 5 rocket launching it blew up on June 4, 1996. The mission, however, had a second chance at success. Spare parts from the first craft were used to build a second. *Cluster* consists of four identical craft, each one named after a dance: *Salsa*, *Samba*, *Tango*, and *Rumba*. They were launched in pairs, this time not from Guiana in South America but from Baikonur, Kazakhstan,

aboard a Soyuz Fregat rocket. The first two lifted off on July 16, 2000 and the second two just three weeks later, on August 9. The four fly in formation, studying the interaction between the solar wind and Earth's magnetosphere.

Ulysses

The *Ulysses* space probe started its mission to the Sun by heading off in the direction of the planet Jupiter, arriving there in February 1992 after a seventeen-month journey. *Ulysses* needed mighty Jupiter's gravity-assist to swing it into its unique solar orbit. The craft's distinctive path allows it to make the first measurements of the unexplored regions above the Sun's poles. It passed over the Sun's south pole first, in 1994, and then over its north pole the following year. It then headed toward Jupiter once again on its six-year orbit of the Sun before passing over the poles for a second time in the years 2000–01. The mission was extended in 2004 so

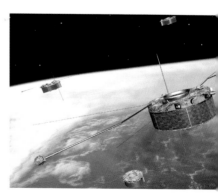

■ An artist's impression of *Cluster*'s four drum-shaped craft flying over Earth.

Ulysses **undergoes pre-flight checks.
The dish antenna is for communications.**

that *Ulysses* will fly over each of the poles for a third time, during 2007–08.

Because the Sun follows an eleven-year-long activity cycle, conditions at the Sun change from one polar pass by *Ulysses* to the next polar pass. On the first pass, in 1994–95, the Sun was at solar minimum—the quiet period of its activity cycle. The second pass spanned the peak of activity—solar maximum. It will be solar minimum once again by the time of the third pass, but there will be an additional change then. The Sun's north and south poles swapped places at the time of solar maximum, so on this third pass the poles will be in opposite positions to those experienced at solar minimum in 1994–95.

Ulysses is a joint venture between Europe and the United States. It was built in Germany and then shipped to Florida, from where it was launched by space shuttle on October 6, 1990. From its unique viewpoint, *Ulysses* has shown the Sun's variable effect on the space surrounding it, from the equator to the poles, and improved our knowledge of the makeup and origin of the solar wind, as well as our understanding of interplanetary and interstellar dust.

SOHO

Like *Ulysses*, *SOHO*, which is short for Solar and Heliospheric Observatory, is a joint European-U.S. project. In fact, it is a truly international affair because about 1,500 scientists from around the world have been involved in the mission. Companies based in fourteen different European countries built the craft. The twelve scientific instruments, from countries including Finland, Switzerland, and England, were assembled in the U.K., and then mated with the service module in France, where it had been assembled. *SOHO* was launched on an Atlas rocket from Cape Canaveral on December 2, 1995.

In a halo orbit

SOHO investigates the Sun's interior, its photosphere, and its atmosphere, as well as the production of the solar wind. *SOHO* operates from a special position in space known as Lagrangian point L_1 (*see pp. 92–93*); it is on the sunward side of Earth, at a site where the gravitational fields of Earth and the Sun cancel each other out. *SOHO* was the first spacecraft to be placed into this type of orbit, known as a halo orbit (*see p. 92*). From here it watches and records the Sun twenty-four hours a day. It has provided the most detailed database of solar surface features, and has returned spectacular images of surface activity, of explosions, jets, flares, and prominences. The craft was intended to operate only until 1998, but its mission has been extended until 2007, which means it will work for one complete solar activity cycle of eleven years.

Sungrazers

One of the craft's instruments routinely monitors a huge volume of space around the Sun. Comets have unexpectedly flown into this region and then been noticed. These new discoveries account for almost half of all comet discoveries. About 85 percent of the comets seen are termed sungrazers, comets that follow paths that take them close to the Sun, within about 500,000 miles (800,000 km) of its visible surface. Before *SOHO*, spacecraft had discovered only sixteen such comets. On August 12, 2002, *SOHO*'s tally reached 500; and by August 2005 it was 1,000. The comets are detected in images of the Sun's outer atmosphere, its corona. The light of the Sun is blocked, so the faint corona, and occasionally a comet, is visible. Amateur astronomers who look through the *SOHO* images on the Internet discover many of the comets.

Genesis

The *Genesis* spacecraft left Earth on August 8, 2001, on its ambitious mission to collect solar wind particles and return these to Earth. This U.S. probe spent the next three months traveling toward the Sun and then moved into its orbital position at the first Lagrangian point. Here it collected solar wind particles in hexagonal collectors. The solar wind consists of tiny particles that have been shot out of the upper layers of the Sun. It is of interest to scientists because it represents the material that was in the solar nebula 5,000 million years ago and subsequently formed into the Sun and its system. It is believed that the material has not been modified by nuclear reactions in the Sun's core.

Genesis's precious cargo was stored and returned to Earth in a capsule. Although the container crash-landed and was damaged, with most of the outer layers of collectors broken and contaminated, some of the collectors survived the journey intact. The first of the captured solar wind particles were sent to researchers to start their scientific study in early 2005.

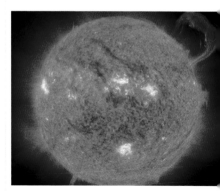

■ **A huge prominence arches out of the Sun in this awe-inspiring image from *SOHO*.**

CHAPTER 5

Missions to Mars

Mars is the planet that most closely resembles Earth. About half Earth's size, Mars is made of rock and metal and has two polar ice caps. It is now a cold, seemingly barren world, but its surface, which has been shaped by volcanism and water, would have been quite different in the past. Mars has always intrigued astronomers because it is the planet most likely to support life. A number of the spacecraft sent to it have tested their landing sites for any indications of life. None has been found, but scientists continue to search and at the same time learn about the conditions in which life on Earth could have started.

Much-visited planet

More than thirty craft have made successful missions to Mars, from the early 1960s to the present. Craft have flown by the planet, orbited around it, landed on it, and driven over it. Half of these have been sent by the United States. Nearly all the rest came from the Soviet Union, with, as of 2006, one from Japan and one from Europe. The first craft intended for Mars was the Soviet Union's *Marsnik 1*. Its aim was to fly by the planet, but its launch in October 1960 was unsuccessful. Twenty-two further probes attempted to fly by, orbit, or land between 1960 and the mid-1970s, when the U.S. *Vikings 1* and *2* touched down on Mars. These probes looked for signs of life,

but their findings were inconclusive. The general feeling of disappointment, coupled with a change of priorities, brought to an end this initial phase of Martian exploration.

Second-phase exploration

Just three craft were sent toward Mars during the twenty years after the Vikings. These were the Soviet Union's ambitious craft *Phobos 1* and *Phobos 2*, each of which consisted of an orbiter and a lander to touch down on Mars's moon, Phobos. They were launched in July 1988 but contact was lost with both. The same fate befell the third craft, the U.S. orbiter *Mars Observer*, in 1993.

The U.S. *Mars Global Surveyor*, launched in 1996, was the first of the second wave of probes to Mars. It was an unprecedented success, along with *Mars Pathfinder*, *2001 Mars Odyssey*, and the Mars Exploration Rovers, called *Spirit* and *Opportunity*, as well as Europe's first mission to Mars, *Mars Express*. There were others, but

◄ *Mars Reconnaissance Orbiter* burns its rocket over Mars's southern hemisphere.

▲ The mighty Proton rocket with the *Phobos 1* probe on board at Baikonur Cosmodrome.

⊡ The giant volcano Olympus Mons (lower center) and the northern ice cap (top).

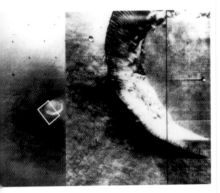

⊡ *Mariner 9*'s wide-angle (left) and telephoto (right) lenses view the crater of a volcano.

these proved to be partial or total failures. The Russians failed to get their orbiter, *Mars 96*, beyond Earth. Japan's only probe to Mars thus far, *Nozomi*, suffered a shortage of fuel after a malfunction early in its mission; rather than go into orbit around Mars in 1999 as scheduled, it eventually flew past the planet in 2003. In March 1999, the U.S. *Mars Climate Orbiter* reached Mars, but contact was lost as the craft moved into orbit; it was set

on the wrong course when commands were sent in imperial rather than metric measurements. Designed to work in tandem with *Mars Climate Orbiter*, *Mars Polar Lander* was intended to touch down near the south polar cap and dig into the Martian surface. It too reached Mars, but, for an unknown reason, contact was lost before the planned landing.

Mariner

The first U.S. craft to Mars were in the Mariner series. *Mariner 4* made the first successful flyby. It was launched on November 28, 1964, just a couple of weeks after *Mariner 3*. The two were to fly by the planet together, but *Mariner 3* could not deploy its solar panels because the protective cover did not release. Consequently the probe could not generate any electric power. *Mariner 4*, however, arrived at Mars on July 14, 1965, and took twenty-two images of the surface, the first close-up views of another planet's surface.

Increasing success

A second joint mission, this time performed by *Mariners 6* and *7*, was launched four years later. *Mariner 6* blasted off for Mars on February 25, 1969, then *Mariner 7* one month later. Between them they took about 200 photographs of the planet and investigated its atmosphere. The last of the Mariner missions to Mars was *Mariner 9*. It became the first craft from any nation to orbit another planet. It moved into orbit in mid-November

1971, but a dust storm on the planet delayed its imaging program. Normal mapping started the following month and more than 7,000 images were returned. They provided the first global surface map, and showed the planet's massive volcanoes and extensive canyon system, Valles Marineris, which takes its name from the *Mariner 9* craft. Images of Mars's moons, Phobos and Deimos, were also taken during the mission.

Soviet attempts

Sixteen Soviet craft consisting of flyby probes, orbiters, and landers were prepared for Mars between 1960 and 1973. Many of them failed and others achieved only partial success. Communication was lost with *Mars 1* in March 1963 as it flew toward the planet. *Mars 2* orbited around Mars, but its descent module, the first to attempt a soft landing on the planet, crashed on the surface. A landing was achieved just days later by *Mars 3* in December 1971, but twenty seconds after landing, its instruments failed. Four more craft in the same series flew to Mars, but all developed problems and returned only limited useful data.

Viking

The two Viking missions, which were the first U.S. craft to land on Mars, each consisted of an orbiter and a lander. *Viking 1* lifted off first, on August 20, 1975. *Viking 2* followed, just three weeks later, on September 9. Their aim was to image the planet's surface,

study the atmosphere and surface, and search for signs of life. On arrival, both craft orbited Mars and returned images that scientists on Earth used to select touchdown sites for the Viking landers. The *Viking 1* lander separated from its orbiter and touched down at Chryse Planitia on July 20, 1976. After orbiting the planet for almost six weeks the *Viking 2* craft separated and its lander touched down in Utopia Planitia on September 3, 1976.

Imaging Mars

As these two worked on the ground, the two orbiters traveled around Mars imaging the planet's entire surface and acting as communication relays for the landers. *Viking 1* made more than 1,400 orbits, completing its work in August 1980. *Viking 2* was turned off on July 25, 1978 after returning almost 16,000 images taken during 706 orbits. The images showed volcanoes, areas flooded by lava, vast canyons, impact craters, and evidence of wind erosion.

◪ *Viking 1*'s camera captures the rock-strewn Martian surface and part of the lander.

The *Viking 1* orbiter had a close encounter with Phobos, the larger of Mars's two small moons, and the *Viking 2* orbiter flew by the other, more distant, Deimos.

Landing on Mars

The Viking landers were six-sided aluminum structures supported by three extended legs. Power for each craft's instruments was generated from onboard decaying plutonium. The instruments included two 360-degree cameras; a sampler arm with a collector head, temperature sensor, and magnet at its end; meteorology instruments; sensors; and finally spectrometers. The sampler arm, controlled from Earth, collected soil. An aeroshell protected and slowed each lander as it approached the Martian surface. Then a parachute and retro-rockets slowed the craft further. *Viking 1* sent the first image of its landing site twenty-five seconds after landing. It operated almost faultlessly for a period of just over six years until November 13, 1982, when contact was lost after a faulty command was sent from Earth. *Viking 2* was turned off on April 11, 1980, after its batteries had failed. Between them they transmitted more than 1,400 images of the two landing sites, studied the atmosphere, took surface samples and analyzed their composition, and looked for signs of life. No evidence of life was found at either landing site.

Mars Global Surveyor

The U.S. *Mars Global Surveyor* was a successful orbiter planned as a rapid, low-cost probe designed to achieve some of the objectives of the lost *Mars Observer* mission. Its initial aim was to acquire data for about five years, until 2002. It should have run out of fuel in early 2003, but a change to its working methods cut its fuel consumption. *Global Surveyor* has almost doubled its original lifespan and is expected to work until at least September 2006. It is the most successful U.S. craft to orbit Mars.

Orbiting Mars

Mars Global Surveyor was launched on November 7, 1996 and arrived at the planet in September of the following year. It spent the next sixteen months moving into a near circular two-hour polar orbit. Here it works at an average height of 235 miles (378 km) above Mars, achieving complete coverage of the planet every seven days. Its long working life has enabled scientists to make an unprecedented study of the

■ *Mars Global Surveyor* has outdone all the previous U.S. Mars missions combined.

changes to the Martian landscape over time. It has studied the surface structure and geological features; in particular it recorded features formed by liquid water, and its data has been used to identify sites for landing craft. Mars's atmosphere and weather patterns are studied, as well as phenomena such as dust storms and dust devils, water clouds and carbon dioxide clouds, and surface frost. So far it has returned more than 250,000 images of the Martian surface.

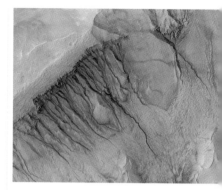

Gullies in the wall of an impact crater are evidence that water once flowed on Mars.

2001 Mars Odyssey

Like *Global Surveyor*, the U.S. *2001 Mars Odyssey* is a box-shaped orbiter with solar panel wings. It was launched in April 2001 and reached Mars in October that same year. After achieving its intended two-hour polar orbit in January 2002, it started its scientific work. As scheduled, it collected data up to July 2004, after which its mission was extended for another Martian year, until September 2006. The craft (named after *2001: A Space Odyssey*, the Arthur C. Clarke novel and Stanley Kubrick movie) is furthering our knowledge of Mars's climatic and geological history, and gathering data to help determine if Mars was ever conducive to life. It is mapping elements and minerals on the surface and locating subsurface water. It has also been looking for radiation that could harm future astronaut explorers, and it has been acting as a communication relay for the two Mars Exploration Rovers, *Spirit* and

This image of *Mars Odyssey* was captured by the camera of *Mars Global Surveyor*.

Opportunity, which have been working on the Martian surface since January 2004. Findings sent by the rovers to *Mars Odyssey* are relayed back to Earth, which receives each radio signal some fifteen minutes later.

Mars Express

Europe's first venture to Mars, and the first fully European mission to any planet, was launched aboard a Soyuz-Fregat rocket from Baikonur,

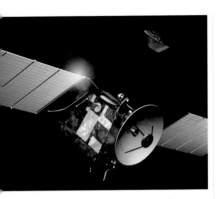

🔲 *Mars Express* releases the capsule containing the *Beagle 2* landing craft (top).

🔲 On Mars, *Beagle 2* was designed to unfold and gain power from its four solar panels.

Kazakhstan, on June 2, 2003. *Mars Express* got its name from its rapid development time; it was built more quickly than any other comparable planetary mission. It is a cube-shaped orbiter, with two solar wings that measure 39 feet (12 m) from tip to tip. At launch it carried a saucer-shaped British lander called *Beagle 2*, named after the ship in which the naturalist Charles Darwin sailed.

Release of *Beagle 2*

After a journey of 250 million miles (400 million kilometers), *Mars Express* arrived at Mars in December 2003; *Beagle 2* was released on December 19 to cruise toward the Martian surface. It entered the atmosphere on December 25 in readiness for landing, but after that its fate is uncertain. No signals were received from *Beagle 2* and the lander was declared lost. *Beagle 2* was to study its landing site and the atmosphere, and search for possible signs of life. Meanwhile the mother craft moved into orbit around Mars, and in early 2004 *Mars Express* started its science operations.

Shared technology

Mars Express uses some technology from *Mars 96*, the failed Russian mission, and aims to fulfill some of that craft's lost scientific goals; other technology came from the European cometary mission, *Rosetta*. The probe is working to answer questions about Mars's geology, atmosphere, surface environment, potential for life, and watery past. It has been mapping the surface geology, mineralogy, atmospheric composition, and subsurface structure. It is also contributing to our understanding of the planet's weather and the evolution of its climate.

Radio-wave investigations

There are seven main instruments on board; these include the high-resolution stereoscopic camera, which has been returning breathtaking images of the surface, and three spectrometers. Protruding from the craft are three antennae booms, which together use radar to look, for the very first time, at what lies below the Martian surface. The craft sends a coded stream of radio waves to Mars at night, and scientists then analyze the waves' distinctive echoes to make deductions about the surface and subsurface structure. On one side of the craft's box-shaped body, an antenna 6 feet (1.8 m) long keeps the craft in touch with the European Space Operations Control Center in Darmstadt, Germany, via its ground station in Perth, Australia. *Mars Express* was to work for one Martian year, until late 2005, but the mission was then extended by another Martian year, about twenty-three Earth months.

Mars Reconnaissance Orbiter

The most recent probe to Mars is the U.S. *Mars Reconnaissance Orbiter*, which blasted off from Kennedy Space Center on August 12, 2005. Its aim is to orbit Mars for a full Martian year, from November 2006 to November 2008, collecting data on its present climate, locating water-related landforms and sites of water activity, and identifying sites for future landing craft, including craft to collect samples for return to Earth. On board are radar instruments for viewing beneath the surface; a sounder for mapping the planet's temperatures, clouds, and dust; and the most advanced cameras yet sent to another planet. Previous cameras to Mars have been unable to identify objects smaller than a bus; one of this orbiter's cameras will be able to see table-sized objects. The mission aims to shed light on how Mars, which was once warm and wet, became the cold, dry place it is today. It will study the history of water on Mars by, for example, looking for deposits of minerals that form in water over long periods of time, and by locating the shorelines of ancient seas and lakes. The craft will also image potential landing sites for future landers.

Aerobraking

Mars Reconnaissance Orbiter will move into orbit around Mars in March 2006 and then spend about eight months aerobraking—dipping into the planet's atmosphere to slow itself down and settle into its final orbit around Mars. Two years of scientific investigation, from November 2006 to November 2008, will be followed by two years

◪ *Mars Reconnaissance Orbiter* uses its climate sounder to map Mars's weather.

acting as a relay station, passing on communications for landing missions. Because *Mars Reconnaissance Orbiter* has fuel for a further five years, this part of the mission can be extended.

Mars Pathfinder

The Ares Vallis region in the northern hemisphere of Mars was the landing site for the U.S. *Mars Pathfinder* mission in 1997. Ares Vallis is a large outflow channel, the site of an enormous flood

⊡ Engineers test air bags designed to protect *Mars Pathfinder* during its Mars landing.

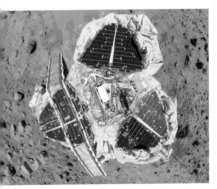

⊡ A probe's-eye view of *Pathfinder* opened amid its deflated air bags on Mars's surface.

in Mars's past. The mission, which consisted of two parts, a stationary lander and a six-wheeled rover, was to investigate the Martian environment before further exploratory missions. The two craft started their mission aboard a Delta rocket on December 4, 1996 and were soon to demonstrate how to make a low-cost landing and exploration of the Martian surface.

Immediate landing

The combined craft entered the Martian atmosphere on July 4, 1997 and prepared to land without going into orbit around the planet. The heat shield and then a parachute slowed the craft, and just before arrival the craft's three retro-rockets were fired to slow the descent further. Four air bags inflated about ten seconds before landing and together formed a 17-foot (5.2-m) protective ball around *Mars Pathfinder*. After touching the ground the craft bounced about fifteen times and rolled across the surface before settling about 1,100 yards (1 km) from its initial landing site.

Deployment of *Sojourner*

Once still, the air bags deflated, and some one and a half hours after landing, the craft opened its three triangular solar panels. Contact was made with Earth, and data collected on the atmosphere during the landing procedure were transmitted home, along with images of the craft and its surrounding area. Once an air bag was moved out of the way, and the ramps placed in position, it was time for the rover to start work. On July 6, the rover, named *Sojourner*, rolled out of the lander on to the Martian surface. *Sojourner*, which was the size of a microwave oven, was controlled from Earth. Images taken by the lander and *Sojourner* helped the operator decide where the rover should be sent. The maneuvers were complicated by a communications delay of ten to fifteen minutes. Messages were sent to the rover's onboard computer via the lander, and any data it collected followed the same route. The rover's power was provided by solar cells affixed to its upper surface; a battery provided supplementary power.

Lander–rover communication

Sojourner took black-and-white images of the lander, the surrounding terrain, and its own wheel tracks in order to increase our understanding of the rocks and soil. A spectrometer on board *Sojourner* assessed the composition of samples taken from sites more than 30 feet (10 m) from the lander. The landing craft, which has since been named the *Sagan Memorial Station*, supported the rover by imaging it and relaying data. The mission ended on September 27, 1997, when communications were lost.

Mars Exploration Rovers

In midsummer 2003, two identical Mars Exploration Rovers, named *Spirit* and *Opportunity*, set out on a mission that would take them roving across the

◪ *Sojourner* rests on *Pathfinder* before opening itself and moving off the lander.

Martian surface. The two carried scientific instruments to investigate their landing sites and were designed to travel about 110 yards (100 m) on each day of their ninety-day assignment. They proved so successful that their mission was extended time and again beyond the original deadline of April 2004. The rovers are now well beyond their original design life and have covered many times the original distance expected of them. As long as no part wears out in the meantime, scientists aim to work them hard until at least November 2006.

Looking for water

The two rovers traveled separately to Mars. *Spirit* and *Opportunity* were launched on June 10 and July 8, 2003 respectively; they landed on the planet half a year later, in January 2004. Their aim was to search for signs of past water activity on Mars and for clues about what the environment was like when the liquid water was present,

An artist's impression of *Spirit*'s robotic arm (left) investigating a Martian rock.

Opportunity was protected by this heat shield during its descent toward Mars.

and whether those conditions could have supported life. There is no liquid water on Mars today, but the rovers soon found evidence that it had been there in the past. Their two landing sites, on opposite sides of the planet, were chosen as likely sites of past water activity. Gusev Crater, where *Spirit* is exploring, is an impact crater that once may have held a lake. *Opportunity* is operating on a plain called Meridiani Planum, which contains hematite. On Earth hematite, an iron-rich mineral, almost always forms in the presence of liquid water.

Landing on Mars

Each of the rovers arrived inside a tetrahedron-shaped landing platform, which in turn was encased in a protective aeroshell. The aeroshell and a parachute slowed the craft as they descended to the Martian surface. Retro-rockets slowed them further just before impact, and air bags inflated to cushion the landing. The landing craft bounced and rolled to a stop, and once the air bags were deflated and retracted, each opened its petal-like sides and the rovers were readied to leave their landing craft.

Geological probing

The two rovers behave like roving geologists, moving across their target areas and stopping from time to time to carry out on-site investigation. They are controlled from Earth by space scientists who perfected their guidance techniques on a rover called *FIDO* in a remote desert location on Earth. The two craft are identical. A computer is housed on a boxlike chassis mounted on six wheels. On top of this is a triangular equipment deck, which has two solar arrays level with it. The solar arrays provide the power, which is stored in two rechargeable batteries. A mast with camera and communication antennae stands on top of the deck, and an arm protrudes from the chassis and out of the front of the rover. At 5 feet (1.5 m) tall, the mast-mounted cameras are the rovers' highest point and provide views of the surrounding area from a vantage point similar to that of a standing human.

Targeted investigations

Images taken by the rovers are used by scientists to identify features such as interesting landforms, rocks, and minerals, particularly those that can form only in the presence of water. The rovers can then be sent to investigate the features; they move along at a top speed of 2 inches (5 cm) per second.

Immediate analysis

Each rover is equipped to analyze rocks and soils *in situ*. The robotic arm has an elbow and wrist and is able to place its instruments up against the rock and soil samples. A microscopic camera in the arm's fist takes high-resolution images. Spectrometers identify the chemical makeup of the rocks, and a rock abrasion tool—called the "rat"—scrapes away the outer layers of a rock to expose its interior for examination. Magnets on the front of the equipment deck collect magnetic dust particles.

Evidence of water

The two rovers very quickly found evidence of Mars's watery past. *Opportunity*, for example, spent its first eight weeks exploring its landing site, a small impact crater named Eagle Crater, within Meridiani Planum. Rock only inches deep within the crater shows the effects of water and indicates that long ago this site was covered in salty water deep enough to create a splash. *Opportunity* then took the half-mile (800-m) journey to Endurance Crater, about the size of a football field, where it worked from May to December 2004. This rover also discovered evidence of Mars's wet past; here, the effects of water were found to be yards deep.

⬆ The slopes of Endurance Crater, where *Opportunity* performed geological analyses.

⬆ *Opportunity*'s rock abrasion tool, covered in red dust after grinding into solid rock.

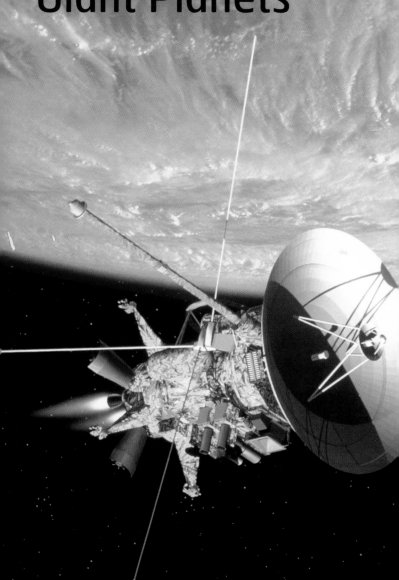

CHAPTER 6

Voyages to the Giant Planets

Jupiter, the largest planet in the solar system, is on average more than five times more distant from the Sun than Earth at about 483.7 million miles (778.4 million kilometers). It is the closest to us of the four largest planets, which are commonly called the gas giants because of the vast gaseous atmosphere that makes up a significant part of each of them. Next distant is Saturn, which has the most extensive ring system of all four. It orbits on average at about 886.5 million miles (1.4 billion kilometers) distant. Uranus is twice as far away again, and Neptune, the most distant of the four, is about 2.8 billion miles (4.5 billion kilometers) from the Sun, thirty times as far as Earth.

Distant destinations

The great distance of these planets meant they had to wait their turn to be explored. Once probes had made successful missions to the near planets, attention was turned to the more distant ones. The first probes to them were the Pioneer craft of the 1970s. *Pioneer 10* flew by Jupiter in 1973; *Pioneer 11* flew to Jupiter in 1974, continuing on to reach Saturn in 1979. The distance of these planets means lengthy gaps between the launch of probes and their arrival. Cassini-Huygens, for example, set off toward Saturn in 1997 but arrived only in 2004, after a journey of seven years. The two most distant of the gas giants,

Uranus and Neptune, have each been visited by only one spacecraft—*Voyager 2* in the 1980s—and there was a three-year gap between the flyby of Uranus and that of Neptune.

Rare visitors

A total of six craft have been to the gas giants. Four of these, the two Pioneers and the two Voyagers, were flybys; the other two were orbiters sent to study Jupiter and Saturn and their large families of moons. *Galileo* made an in-depth study of Jupiter and its system in the 1990s, and *Cassini-Huygens* investigated Saturn, with its rings and moons, in the first decade of the twenty-first century. Comprehensive studies such as these continue for years at a time.

Dramatic discoveries

Six craft is a small total compared with the tens of craft sent to the rock planets—Mercury, Venus, and Mars. But each of the six was ambitious for

◧ The *Huygens* probe enters Titan's atmosphere after release by *Cassini*.

◰ *Pioneer 10* undergoes pre-launch testing. Its communication dish is seen in profile.

its time, and proved to be a triumph of technology. The probes opened up the outer solar system to us, revealing the turbulent atmosphere of Jupiter and the complexity of Saturn's rings, and showing Uranus and Neptune for the first time; these planets are just too far distant to be seen in any detail from Earth. They also revealed the larger moons of the outer planets as worlds in their own right. Most are frozen ice worlds, but one, Jupiter's moon Io, was

🖼 Jupiter was photographed by *Pioneer 10* as it flew past in December 1974.

🖼 Saturn, its rings, and its largest moon Titan (top) as imaged by *Pioneer 11*.

a great surprise. This red, yellow, and brown moon was the first world beyond Earth found to have volcanoes that are active.

Surviving the Asteroid Belt

Any craft traveling to the gas giants has first to pass through the Asteroid Belt, beyond the orbit of Mars. This doughnut-shaped region, which consists of billions of chunks of rock, is more than 186 million miles (300 million kilometers) wide. Typically, thousands of miles lie between individual asteroids, but there was a collective sigh of relief among space scientists and astronomers when the first craft successfully negotiated the hazards of the belt.

Pioneer 10

On March 2, 1972 *Pioneer 10* was launched, its mission being to travel through and beyond the belt and become the first craft to reach Jupiter. *Pioneer 10* was the fastest man-made object to leave Earth when its Atlas-Centaur rocket sent the craft off at 32,400 mph (51,800 k/h)—fast enough to cross Mars's orbit in just twelve weeks. The probe entered the Asteroid Belt in mid-July 1972, and flew by Jupiter on December 3, 1973.

Pioneer 11

Even before *Pioneer 10* had arrived at the planet, a second craft in the Pioneer series was launched to Jupiter on April 5, 1973. *Pioneer 11* also crossed the Asteroid Belt safely, flying by

Jupiter on December 2, 1974 and then moving on to Saturn for a flyby on September 1, 1979.

Pioneer equipment

The dominating feature of the identical craft was a dish-shaped communications antenna, 9 feet (2.7 m) wide, fixed to the hexagonal equipment compartment. Each craft carried more than ten experiments for investigating Jupiter, its satellites, interplanetary dust particles, and the solar wind. A plaque with drawings of a man and a woman, and showing the location of Earth in the solar system and galaxy, was mounted on the spacecraft to show its place of origin to any being that might come across the craft in the future.

Long-lasting missions

The Pioneer craft provided the first close-up images of Jupiter and Saturn, of their moons and of Saturn's rings. They collected information on Jupiter's magnetic field, radiation belts, and atmosphere that proved vital when designing the next craft to Jupiter, the Voyager and Galileo missions.

After leaving Jupiter and Saturn behind, the Pioneer craft studied the solar wind and cosmic rays. One by one, over the course of the following years, *Pioneer 10*'s instruments failed or were turned off to conserve power. Its science mission finally ended on March 31, 1997. *Pioneer 11*'s mission had stopped already. By September 1995 none of its instruments was operating; the last communication

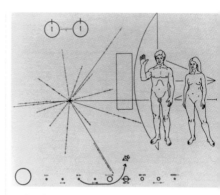

■ The Pioneer plaque includes humans and a simple map of the solar system (bottom).

with the craft was in November of that year. It is today headed toward the constellation of Aquila. The last signal from *Pioneer 10* was detected on January 23, 2003, more than thirty years after its launch, and although further attempts were made to contact the craft, it was concluded that it had run out of power. *Pioneer 10* is heading out of the solar system, in the general direction of the star Aldebaran in the constellation Taurus. It has a long way to go—it would take about two million years to get there.

Planning a grand tour

In the mid-1960s a U.S. space agency student worked on a schedule for future missions based on the positions of the planets in the years ahead. He computed gravity-assist missions whereby a single craft could use the gravity of a planet to accelerate it and redirect it on to another. The student found that the positions of the gas giants in the late 1970s and 1980s would allow a craft to fly by all four planets in a twelve-year mission. At that time the planets were on the same side of the Sun and positioned in an arc, with each planet slightly ahead of the inner one. This meant a craft could use gravity-assist from one to the next, and reduce the flying time to Neptune from thirty to twelve years. Such a planetary alignment occurs only once every 175 years, and so it was a great opportunity to visit the gas giants—one that could not be missed. The next chance would be in 2155. The mission was to be made by two identical craft and soon became known in planetary exploration as the Grand Tour.

Voyagers 1 and 2

The craft, *Voyagers 1* and *2*, were launched aboard Titan-Centaur rockets from Cape Canaveral, Florida, in 1977. *Voyager 2* was launched first, on August 20. It eventually flew by all four planets: Jupiter, Saturn, Uranus, and Neptune. *Voyager 1* set off three weeks

🔲 The U.S. Jet Propulsion Laboratory control room as *Voyager 1* encounters Saturn.

later, on September 5, heading for Jupiter and then Saturn. It was the second to start but the first to reach a target because it was on a shorter, faster flight path; it overtook *Voyager 2* in the Asteroid Belt. Radio instructions from Earth kept the craft on course. Guidance systems on board established their positions in relation to the Sun and a bright star, such as Canopus, as they flew. Small rocket thrusters would fire to correct any misalignments with the Sun and star and place the craft back on course.

Extended missions

When originally conceived, the Voyager craft were intended to last five years and to investigate just two planets, Jupiter and Saturn. This was to be a two-craft, two-planet mission. It was thought too costly to build craft that had to survive the long journey to Uranus and then Neptune. It was only after launch, and once *Voyager 2* had proved to be a success as it flew by Jupiter and later Saturn, that the decision was taken to extend its flight to Uranus. Money was put aside to fund the administration of this leg of the journey and for *Voyager 1* to continue operating too, making a study of interplanetary space as it flew through the solar system. Later, additional funds were found to extend *Voyager 2*'s mission to Neptune.

The most noticeable part of a Voyager craft was its dish-shaped antenna for communications, larger than that of the Pioneer at 12 feet

(3.6 m) wide. The dish was mounted on a ten-sided framework housing the computer command system. Most of the instruments were at the end of a giraffelike neck protruding from the frame. These included a spectrometer and a light detector, as well as a steerable scan platform that held wide-angle, narrow-angle, and infrared cameras. Infrared and ultraviolet sensors, magnetometers, and other sensors and detectors were also mounted on the craft.

◩ Jupiter's Great Red Spot (top), imaged by *Voyager 1*, is a storm larger than Earth.

Nuclear-powered craft

Spacecraft traveling to the gas giants go too far from the Sun to use solar panels to generate the electricity needed to power their equipment, such as computers, navigation systems, collecting and recording instruments, and the antennae to communicate with Earth. Their alternative is nuclear energy; the electricity is generated from the heat produced by the natural decay of onboard plutonium.

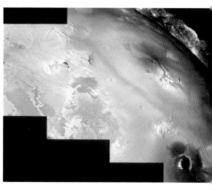

◩ On Io, *Voyager 1* found the Solar System's first active volcanic site away from Earth.

Jupiter and Saturn

Voyagers 1 and *2* flew by Jupiter in 1979. *Voyager 1* made its closest approach on March 5, and *Voyager 2* on July 9. Both craft turned their cameras on to the colorful bands, zones of gaseous atmosphere that encircle the planet. The craft identified storms and eddies; the largest, the Great Red Spot, was found to be a giant storm that spins counterclockwise every six to seven days. Jupiter's faint dusty ring was discovered, as well as three small moons, Metis, Adrastea, and Thebe, between Jupiter and Io. The Voyagers also turned their attention to Jupiter's four largest moons. They found that Io is volcanically active; nine volcanoes were seen to erupt. By contrast, Ganymede, Callisto, and Europa were found to have icy crusts.

Two years later, the two craft arrived at Saturn; *Voyager 1* reached it on November 12, 1980, and, nine months later, *Voyager 2* made its

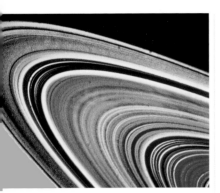

■ A color-enhanced *Voyager 2* image shows the many ringlets in Saturn's ring system.

closest approach on August 25, 1981. Oval storms and clouds were found in the planet's muted atmospheric bands, and new rings, both thin and broad, were discovered. The rings were also found to be made of thousands of ringlets, which in turn are made of dust-sized to house-sized pieces of material. Kinks and spokes were seen in the rings, and shepherding moons that confine particles within the rings.

Uranus and Neptune

Voyager 2, still the only space probe to be sent to Uranus, flew by the planet on January 24, 1986. It detected the planet's magnetic field, as then unknown, which is tilted by an unexpected amount to the planet's axis of rotation. Surprisingly, it was found that most of the planet, including the sides facing and pointed away from the Sun, is at the same temperature—about −350 degrees F (−212 degrees C). Ten moons were discovered, to add to the five then known. One of the previously known moons, Miranda, turned out to be a very strange world; its jumbled surface suggested it had been broken apart and reassembled.

The blue world of Neptune was last to be visited by *Voyager 2*; it flew within 3,000 miles (5,000 km) of the planet on August 25, 1989. At this point in its orbit, Neptune was the most distant planet in the solar system; it was farther away from the Sun than Pluto, the planet that is usually the outermost. On each of its 248-year orbits, Pluto comes inside Neptune's orbit for a twenty-year period. Pluto returned to its role as the most distant planet in 1999, when its orbit re-crossed that of Neptune.

Voyager 2 observed bright streaks of cloud moving in Neptune's upper atmosphere. An irregularly shaped one that whizzed around the planet every sixteen hours or so was nicknamed the Scooter. There were also several dark spots; the biggest, named the Great Dark Spot, was about the size of Earth. The winds were found to be particularly strong, blowing at up to 1,200 mph (2,000 km/h) and stronger than on any other planet.

Rings and moons

Before *Voyager 2* arrived astronomers had identified arcs of material around Neptune. The probe revealed that Neptune has complete, albeit very diffuse rings. Six moons were also discovered, and when *Voyager 2* scrutinized Triton, the largest, already known moon, it found active geyserlike eruptions of nitrogen and dark dust above the frozen surface.

The two craft had together explored all four gas giants, forty-eight of their moons, and the planets' rings and magnetic fields. The information sent back to Earth helped to answer many of the questions astronomers then had about planetary astronomy, and raised new ones about the origin and history of the planets, their rings, and moons.

Onward journeys

The Voyagers' work is not over. The five-year lifetime of the two missions, which then stretched to twelve, continues. The craft remain on the U.S. list of active missions. *Voyager 1*'s path was bent at Saturn so that it could fly by Titan and behind the planet's rings. This set it on a course that took it north and out of the plane (ecliptic plane) in which the planets orbit the Sun. It journeys at the rate of about 320 million miles (520 million kilometers) a year.

Farthest traveler

After its closest approach to Neptune, on August 25, 1989, *Voyager 2* flew south and out of the ecliptic plane. Like *Voyager 1*, it too is heading out of the solar system and for interstellar space but at a slightly slower pace, 290 million miles (470 million kilometers) each year. Both craft are expected to operate until at least 2020, sending back data as they travel; daily communication is made to both craft. The data from the Voyager craft

is collected by ground stations in California's Mojave Desert; near Madrid, Spain; and in Tidbinbilla, near Canberra, Australia. On February 17, 1998, *Voyager 1* overtook *Pioneer 10* to become the most distant craft from Earth in space.

As they travel, the Voyagers collect data on the Sun's magnetic field and the solar wind. Information sent back to Earth has enabled scientists to establish approximately where the

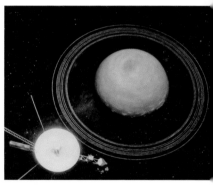

◩ *Voyager 2* flies by Uranus, which appears blue due to methane in its atmosphere.

◩ A *Voyager 2* image of high-altitude clouds above Neptune's southern hemisphere.

There is also video coding for images of Earth showing what it was like in the twentieth century.

■ An impression of *Galileo* releasing its probe (left) into Jupiter's atmosphere.

■ Europa, by *Galileo*. Blue indicates pure ice in the color-enhanced image (right).

solar system ends and interstellar space begins. Called the heliopause, this is the outer boundary of the solar wind. They believe it is between 90 and 120 astronomical units from the Sun (one astronomical unit is the distance between Earth and the Sun). Should an intelligent alien being ever intercept either craft, it will be able to learn much about Earth. Each craft has on board a gold-plated record of sounds of Earth, along with the means to play it.

Galileo

The idea of a new mission to Jupiter took shape almost as soon as the Voyager craft had completed their flyby of the planet. The *Galileo* probe was designed to make a thorough study of Jupiter and its system, orbiting around the planet and flying close to its four major moons in turn. The U.S. craft carried experiments for scientists from around the world, from Britain, Canada, Sweden, Germany, and France, as well as from the United States. *Galileo* was carried into Earth orbit by the space shuttle *Atlantis* on October 18, 1989 and then propelled on to its path to Jupiter. Its course took it by Venus and then Earth twice, for gravity-assist maneuvers whereby it gained energy on each encounter. During its six-year journey to Jupiter, *Galileo* recorded the first close-up images of an asteroid, and it was the only spacecraft with a direct view of the fragments of comet Shoemaker-Levy 9 as they plunged headlong into Jupiter's atmosphere.

Instrumentation

The spacecraft's innovative design consisted of two main sections that were joined together but allowed independent movement. One section, the base, carried cameras. The second, spinning section had instruments for measuring the space environment and magnetic fields; it rotated slowly as it worked. Attached to this section were the communications antennae, the computers, and other support systems. *Galileo* also carried an atmospheric probe with its own seven instruments for studying Jupiter's atmosphere.

Arrival at Jupiter

The probe was released from the main craft when it was 53 million miles (85 million kilometers) from the planet on July 13, 1995. The main craft and probe then cruised independently to Jupiter, arriving at the same time, on December 7, 1995. The wok-shaped probe plunged into the atmosphere and for fifty-eight minutes transmitted information on its temperature, pressure, and composition, and on winds and lightning. It was the first spacecraft to measure Jupiter's atmosphere directly, and to make long-term observations of the system.

Close encounters

Meanwhile, the main craft moved into orbit around the planet to start its tour of twenty-three months and eleven orbits, which was later extended to include twenty-four more orbits. *Galileo* found that Jupiter has thunderstorms with lightning up to 1,000 times more powerful than that on Earth, and revealed that its ring system is formed of dust knocked off its innermost moons. The flybys it made of Jupiter's moons were between 100 and 1,000 times closer than those undertaken by the Voyager craft.

Galileo made eleven close encounters with Europa, getting in as close as 125 miles (200 km), and found evidence of a liquid ocean under its icy surface. It got even closer to Callisto, and to Io, where it recorded changes caused by volcanic eruptions since Voyager had been there.

End of *Galileo*

Not everything went smoothly. When the time came to unfurl the 16-foot (4.8-m) diameter umbrella-style antenna, in April 1991, up to three of its ribs remained closed. Fortunately, *Galileo*'s other antennae were in good working order, although they sent and received data at a slower rate. The mission ended on September 21, 2003, when *Galileo* was deliberately plunged into Jupiter's atmosphere to prevent it from making an unwanted impact with Europa, contaminating a world that could have a subsurface ocean possessing the ingredients for life.

◪ A *Galileo* image of craters on Ganymede, possibly formed by comet fragments.

🔺 *Cassini* releases the *Huygens* probe into the haze of Titan's upper atmosphere.

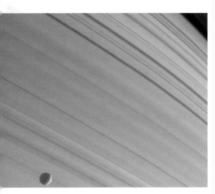

🔺 A *Cassini* image of the icy moon Mimas and the northern hemisphere of Saturn.

Cassini-Huygens

The bus-sized *Cassini-Huygens* mission to Saturn, which was launched on October 15, 1997, is a joint U.S.-European venture. It is a sophisticated craft consisting of two probes, hence its name. The work of *Cassini*, which is the larger of the two and the U.S. part of the mission, is to make an in-depth, four-year study of Saturn, its ring system, and its large number of moons. The smaller European probe, *Huygens*, which was attached to *Cassini* for the journey to Saturn, was released six months after arrival and parachuted to the surface of Titan, Saturn's largest moon. Titan is of interest because this moon has a smoggy atmosphere and may have conditions similar to those found on Earth before life began.

International project

The two together make one of the largest, heaviest, and most complex interplanetary spacecraft ever built. The craft is also a symbol of international collaboration. Hundreds of scientists and engineers from sixteen European countries and the United States have designed, built, and operated the two craft. Italy, for example, contributed *Cassini*'s dish-shaped antenna. The journey to Saturn took seven years and included four gravity-assist maneuvers, two past Venus, and one each by Earth and Jupiter. During the six months that *Cassini-Huygens* was near Jupiter in 2000 to 2001, it worked with the *Galileo* probe. Together they observed and measured Jupiter's magnetosphere and other parts of Jupiter's system, work that neither spacecraft could have achieved alone.

Arrival at Saturn

Cassini-Huygens arrived at Saturn on July 1, 2004, just twenty-three years after the last craft, *Voyager 2*, had flown by. Its main engine was fired to reduce its speed, and the planet's gravity pulled it into orbit around it.

It then proceeded on the first of its seventy-four orbits, which will let it study Saturn's polar regions as well as the planet's equatorial belt. During its four-year mission it will also make forty-four close flybys of Titan, and flybys of other moons.

Commissioned investigations

Cassini's twelve instruments include remote-sensing equipment such as cameras and spectrometers, and direct-sensing instruments to investigate the environment around the spacecraft, looking at, among other things, the quantity and composition of dust particles. In all, the instruments will carry out twenty-seven different science investigations on behalf of 250 scientists worldwide.

Landing on Titan

As *Cassini* made its third orbit of Saturn, *Huygens* was released to start its twenty-one-day journey to Titan. *Huygens* took two and a half hours to descend through the moon's opaque atmosphere to the unseen and unknown surface. Its six instruments tested the nitrogen-rich atmosphere, and then, as the craft broke through the clouds, it imaged the rapidly approaching surface. Astronomers were not sure what the surface would be, perhaps liquid, or solid, or something between the two. On January 14, 2005, *Huygens* touched down on sand made of ice, covered by ice pebbles. *Cassini* continues to study Saturn and its system; it will work until

the predicted end of the mission on July 1, 2008. It has already monitored giant storms in the planet's upper atmosphere, and returned detailed information about the structure and composition of the ring system. It has discovered a new town-sized moon within the rings, and provided a fresh and detailed look at moons already seen by *Voyager 2*, such as Enceladus, as well as close-up images of smaller moons such as Phoebe and Mimas.

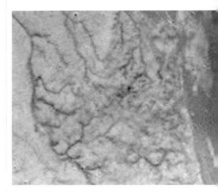

This *Huygens* image of Titan appears to show drainage channels and a shoreline.

Mimas is covered in impact craters; the largest one by far is Herschel Crater.

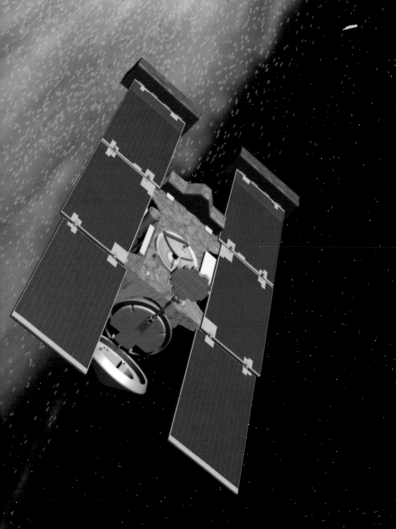

CHAPTER 7

Traveling to Minor Worlds

There are trillions of minor worlds in our solar system. These comprise the asteroids, most of which are found between the orbits of Mars and Jupiter; the Kuiper Belt objects that orbit beyond Neptune; and the comets, the vast majority of which exist way beyond in the spherical Oort Cloud that surrounds the solar system. These minor worlds range in size from dust to irregularly shaped boulder- and city-sized objects, as well as a small number of spherical bodies that are similar in size to Pluto. The asteroids are rocky, or rocky-metallic bodies; farther from the Sun, the Kuiper Belt objects and comets are a mixture of rock and ice.

Comets and asteroids

Astronomers are interested in comets because they are made of material that has remained unchanged since the birth of the solar system, some 4.6 billion years ago. By studying a comet's pristine material we learn more about the origins of Earth and the other planets. Eight craft have so far encountered comets; five of them have visited Halley's comet, and another has collected cometary samples for return to Earth. A ninth, the European craft *Rosetta*, is on its way to land on a comet in 2014.

The study of asteroids can help us understand the period of the solar system's past when Earth and the other planets formed from the coming together of such bodies. A number of craft have flown by asteroids as they traveled on route to other targets, and two craft have been sent specifically to asteroids. The *Galileo* craft—on route to Jupiter—supplied the first-ever close-up image of an asteroid when it flew by Gaspra, 12.5 miles (20 km) long, on October 29, 1991. Almost two years later it flew by Ida, which was discovered to have its own asteroid moon, Dactyl.

NEAR and *Hayabusa*

The U.S. probe *NEAR* (Near Earth Asteroid Rendezvous) made the first landing on an asteroid in 2001, and in November 2005 the Japanese probe *Hayabusa* collected a sample from the asteroid Itokawa for return to Earth. The Kuiper Belt objects and the planet Pluto, which is possibly a large-sized member of this group, are expected to be similar to comets in composition. Space probe exploration should shed

◪ *Stardust* approaches comet Wild 2 with its round sample collector (blue) in position.

◪ More than 600 craters have been counted in this *Galileo* view of Gaspra alone.

light on these worlds and establish how Pluto—planet or minor member—fits into the solar system.

Halley's comet

When Halley's comet returned to the inner solar system and to Earth's sky in the mid-1980s, a great opportunity was presented for sending the first space probe to a comet. Halley's comet was an ideal candidate for the space probe. Its orbit was well known, and

▲ *Giotto* travels toward Halley's comet, at the same time relaying data back to Earth.

▲ The nucleus of Halley's comet produces jets of dust as it is heated by the Sun.

space scientists could accurately pinpoint where to direct their probes. It had also been seen many times before and would be seen again on this return, enabling data from space to be compared with data from ground-based telescopes. A number of countries took this unique opportunity, and a total of five craft were launched to the comet.

Giotto

By far the most ambitious was *Giotto*, sent by the European Space Agency. *Giotto* was launched on July 2, 1985, by a French Ariane rocket from Kourou in French Guiana. The drum-shaped craft, 6 feet (1.8 m) wide, was covered in solar cells to provide the power. It had a dish antenna for transmitting data and receiving instructions from Earth. Ten investigative instruments, including a camera, spectrometers, and dust detectors, were mounted around the drum on the side opposite to the antenna. A bumper shield, unique to the craft, protected the instruments from the hail of dust particles that the probe would experience as it flew instruments-first into the comet.

Toward the nucleus

Giotto's mission was to fly into the coma, the comet's large head of gas and dust, and travel as close as possible to the mountain-sized nucleus within the coma. It was hoped that the craft would get within a few hundred miles of the nucleus, so that its camera could capture the first image of a "dirty-snowball" nucleus. *Giotto* flew to within 370 miles (595 km) of Halley's nucleus on March 13, 1986, and revealed a potato-shaped dirty snowball about 9 miles (15 km) long, with a width and depth of about 4 miles (7 km). *Giotto* continued to fly through the comet, but sent no more images because its camera had been hit by cometary dust and was no longer working. Although other components of the craft had also been damaged, enough were working to make a further cometary encounter viable. Accordingly, *Giotto* was directed to comet Grigg-Skjellerup, and flew within 125 miles (200 km) of this comet on July 10, 1992.

⊡ Halley's comet seen by the *Vega 1* probe. The nucleus is hidden in the large coma.

Other probes to Halley's comet

The other missions to Halley's comet were not so ambitious. None would even attempt to get as close as was planned for *Giotto*. Two Japanese probes, named *Sakigake* and *Suisei* (previously called *Planet A*), left Earth in December 1984 and August 1985 respectively. They both monitored the solar wind near Halley's comet. *Vega 1* and *Vega 2* were sent by the Soviet Union in December 1984; they were similar in design to the Venera craft that the Soviets had sent to Venus. The Vegas flew by the comet on its Sun-facing side.

Although the U.S. *ICE* (International Comet Explorer) mission was not targeted at Halley's comet, it also deserves mention. This craft spent the years 1978–84 probing the Earth's magnetosphere before being diverted to follow an orbit that passed through the tail of comet Giacobini Zinner on September 11, 1985. *ICE* therefore became the first space probe to encounter a comet, as *Giotto* was not scheduled to reach Halley until the following year.

Deep Space 1

The primary aim of the U.S. *Deep Space 1* mission was to test new spacecraft technology. This was achieved within the first few months of the mission's launch on October 24, 1998. The probe was then directed to fly by the asteroid Braille on July 28, 1999, before the mission was extended to include an encounter with comet Borrelly. Although the probe's lifetime was now more than three times its intended duration, the encounter with Borrelly was flawless; the probe passed by the comet's nucleus on September 22, 2001, and transmitted images back to Earth. This was the second cometary nucleus to be seen, and the first to be imaged by an American craft.

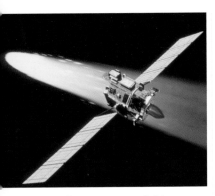

◪ *Deep Space 1* prepares to fly into comet Borrelly's head of gas and dust.

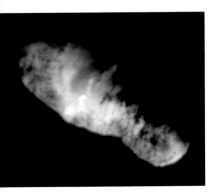

◪ *Deep Space 1* images of comet Borrelly reveal a dry, sooty black nucleus.

CONTOUR

The next U.S. mission to a comet was an ambitious one. *CONTOUR* (Comet Nucleus Tour) was to fly by three comets—Encke, Schwassmann-Wachmann 3, and d'Arrest. The probe was launched on July 3, 2002, but by August 15 all contact with the craft was lost. It is believed that massive structural failure was responsible for the breakup of the craft before it left Earth's orbit.

Deep Impact

The U.S. mission *Deep Impact* to comet Tempel 1, beginning on January 12, 2005, was a flyby operation with a difference. Within six months of its launch the probe had made a crater in the comet's surface and was exploring its interior. The probe consisted of two parts. The main part was the flyby craft, which was a box-shaped framework with a flat rectangular shield to give protection against high-speed dust particles. This carried the second part, the impactor, a battery-powered hexagonal box craft only 3.25 feet (1 m) wide.

Impacting the nucleus

The combined craft flew to the comet, arriving there at the start of July 2005. Images of the nucleus were taken and, with its tracking telescopes pointing at Tempel 1, the main craft released the impactor on July 3. This maneuvered into the path of the comet and impacted the cometary nucleus at about 23,000 mph (37,000 km/h) the following day. A fireball of vaporized impactor and cometary material shot out from the nucleus, which expanded into a fan shape in the hours after impact. This was accompanied by an immense flash of light. A camera on the impactor took images of the nucleus until just three seconds before colliding with it. Meanwhile, the main craft observed and recorded the impact, the material ejected from the nucleus, and the crater itself, which exposed the interior of the comet. Observatories around Earth, and others in orbit above Earth, observed the impact to collect the maximum amount of information from the event. The Hubble Space Telescope and the seven telescopes of the European Southern Observatory in Chile, as well as the *Rosetta* space probe, which was on its way to land on another comet, all captured the impact.

Stardust

Astronomers have always wanted to get their hands on some cometary material and analyze it in their laboratories. A few years ago this was just a dream, but in early 2006 it became a reality. The U.S. *Stardust* mission—the country's first to be dedicated solely to the exploration of a comet—flew to comet Wild 2 and collected dust particles emitted by this comet's nucleus. The probe returned without mishap, and the capsule containing the particles was back on Earth in January 2006.

Collecting cometary dust

Stardust's journey began on February 7, 1999, first flying close to Earth in a gravity-assist maneuver before flying by the asteroid Annefrank in November 2002. The probe arrived at comet Wild 2 on January 2, 2004, flying within 150 miles (240 km). The probe imaged the nucleus, 2.8 miles (4.5 km) wide with bowl-shaped depressions covering its surface, and survived the bulletlike impact of millions of up to pea-sized particles.

The cometary sample was collected during the flyby in a tennis racket-shaped collector, which trapped the particles in an extraordinary substance called aerogel, a spongelike, silica-based solid that is 99.8 percent air. This collector was then retracted into the probe's reentry capsule for the journey back to Earth. The capsule was released as *Stardust* passed by Earth, landing in the Great Salt Lake desert, in the United States, in the early

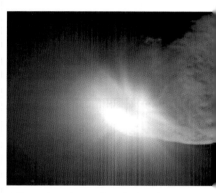

◪ The nucleus of comet Tempel 1 just after it was struck by *Deep Impact*'s impactor.

◪ A scientist displays a tray used by *Stardust* to collect particles of comet Wild 2.

⬈ Scientists open the *Stardust* capsule containing samples of comet Wild 2.

⬈ An impression of the lander *Philae* on the nucleus of comet Churyumov-Gerasimenko.

morning of January 15, 2006. Its safe landing marked the first return of material from outside the Moon's orbit.

Rosetta

The Europeans have already launched their second probe to a comet. This one is even more ambitious than *Giotto*; it consists of an orbiter and a lander, which is destined to make the first touchdown on a cometary surface. *Rosetta* set off on March 2, 2004, in order to travel to comet Churyumov-Gerasimenko but will not rendezvous with its target for another ten years. It will first make three Earth and one Mars gravity-assist maneuvers before encountering the comet. It will move into orbit around the comet's nucleus in May 2014 and then travel with it as it heads into the inner solar system.

Landing on Churyumov-Gerasimenko

When *Rosetta* first encounters the comet it will be far from the Sun and inactive. As the comet approaches the Sun and the Sun heats it up, it will change. *Rosetta* will monitor the change in the comet, examining the formation and then decay of the comet's coma and tails as the comet travels in, and then away from the Sun. In November 2014 the lander, named *Philae*, will touch down and harpoon itself to the surface, take panoramic images, and make an analysis of the nucleus's composition. Meanwhile, the box-shaped orbiter will continue to orbit the cometary nucleus. An added bonus of the mission is that *Rosetta* will fly by two asteroids on its way to comet Churyumov-Gerasimenko. It will first encounter town-sized Steins on September 5, 2008, which measures only 6 miles (10 km) or so across. Next is the larger asteroid Lutetia on July 10, 2010, which is about ten times larger.

NEAR and Eros

The U.S. *NEAR* mission must qualify as the most exciting of all space-probe missions. *NEAR* (later renamed *NEAR*

Shoemaker) was designed to study the asteroid Eros for about a year as it orbited around it. Locating and orbiting such a tiny body—Eros is only 21 miles (33 km) long, and 8 miles (13 km) wide and deep—is an outstanding achievement. But the small-scale craft did much more than this; it made a successful landing on Eros's rocky surface. This was an incredible feat for a craft that was not designed as a lander.

Passing Mathilde

NEAR started its journey at Cape Canaveral, Florida, on February 17, 1996. A Delta rocket launched the ten-sided, drum-shaped craft. Once in space its four solar panels moved into their windmill arrangement. On board were instruments such as photographic equipment and spectrometers for studying Eros. On its way to Eros the probe flew by and imaged the asteroid Mathilde, on June 27, 1997. Then it flew by the Earth for a gravity-assist maneuver to take the craft onward to Eros. Software problems and loss of contact put the mission in jeopardy, but control was eventually regained, and *NEAR* moved into orbit around Eros on February 14, 2000.

Landing on Eros

NEAR remained in orbit until late January 2001, when the craft made a number of low passes, taking it within a couple of miles of the asteroid's surface. A controlled descent to the surface, at the speed of a city bicycle,

followed on February 12, 2001, just one year after arriving at the asteroid. *NEAR* landed on the tips of two solar panels and the bottom edge of its body. The craft continued to work until final contact on February 28; by then it had provided ten times more data than had been planned. *NEAR* had taken about 160,000 images of Eros and in the process had achieved many mission firsts in space exploration. It was the first solar-powered craft to fly beyond

🖼 An artist's impression of *NEAR* flying over Eros' cratered terrain.

🖼 How a rock would roll on Eros: fastest in the red areas, not at all in the blue.

the orbit of Mars, the first to go into orbit around an asteroid, and the first to land on an asteroid.

Hayabusa and Itokawa

The aim of the Japanese probe *Hayabusa* (previously called *Muses-C*) is to fly to the asteroid Itokawa, collect a sample from its surface, and then return this to Earth. *Hayabusa* is a box-shaped craft just 5 feet (1.5 m) long, with two solar panels. It was launched

on May 9, 2003, from the Kagoshima launch center and was put on a path that brought it back to Earth for a gravity-assist flyby one year later. The craft then headed for Itokawa, imaging the asteroid as it approached it in August 2005. On arrival at its target in September 2005 *Hayabusa* did not orbit the asteroid but followed an orbit around the Sun, which kept it close to the asteroid.

Sampling the asteroid

The probe then studied the asteroid's surface from a distance of about 12 miles (20 km) for some weeks before moving in for a soft landing and collection of a sample. On November 20 a target marker was successfully released and this guided the craft to the surface. *Hayabusa* landed and then moved off but it is uncertain whether its funnel-shaped horn collected a sample. *Minerva*, a miniprobe with a camera, was also released but it was too high above Itokawa and it did not make it to the surface.

Returning the capsule

Only days after *Hayabusa* lifted off from Itokawa communications were temporarily lost. The added problems of a fuel leak and propulsion difficulties meant that in early December 2005 the decision was made that *Hayabusa* would not start its journey home for another three years. It was to return to Earth in June 2007 when its reentry capsule would land via parachute near Woomera, in

◪ An impression of *Hayabusa* preparing to land on the surface of Itokawa.

◪ Itokawa has a surprisingly rough, rock-strewn surface with large smooth areas.

Australia. This is now scheduled for June 2010. Japanese space scientists will be interested to see if there is a sample inside the capsule, but the success of the mission this far has opened the way for future Japanese sample-return missions.

Pluto and Kuiper Belt flyby

Before the discovery of Kuiper Belt Object UB313 in 2003, Pluto was believed to be the smallest and most distant planet, being, on average, forty times farther from the Sun than Earth. Pluto and UB313 are unlike their planetary neighbors, and some astronomers question their status. Pluto and its moon Charon more closely resemble other objects in the Kuiper Belt, which exists beyond the orbit of Neptune, out to about fifty times farther from the Sun than Earth.

New Horizons

The U.S. *New Horizons* space probe, which launched from Cape Canaveral on January 19, 2006, will fly by Pluto and Charon and then onward to a number of Kuiper Belt objects. It is hoped that *New Horizons* will solve the question of Pluto's status by revealing its nature, and how this compares with Kuiper Belt objects and comets.

From Jupiter to Pluto

New Horizons has a triangular, cheese-wedge shape; it is topped by a dish antenna and has a plutonium power-source protruding from one point of the triangle. The probe will fly by

■ In Florida, *New Horizons* starts its journey to Pluto aboard an Atlas V rocket.

■ An impression of *New Horizons* approaching icy Pluto and Charon.

Jupiter in February 2007 for a gravity boost and four months of study. Then all but the most critical systems will be turned off as it cruises to Pluto. In June 2015 it will start five months of study as it approaches and then passes Pluto and Charon. It will investigate surface properties, geology, interior makeup, and atmosphere. On leaving Pluto, *New Horizons* will spend the next five to ten years traveling toward and examining Kuiper Belt objects.

CHAPTER 8
Future Exploration

Astronomers and space scientists are always looking ahead. The rapid growth of their knowledge and technological advances means that there are always new questions to be asked, and ways to answer them. At any one time, there are space probes on their way to targets in the solar system, others relaying data from a planet or moon back to Earth, or further craft returning home with, for instance, a tiny sample of a comet. Older space probes, their job done, are silently moving away from Earth and out of the solar system, or endlessly orbiting a planet. New craft to replace them are under construction, and yet another generation are at the design stage, while, looking even further ahead, hopes and dreams for future decades are being worked into realistic plans.

Already on their way

A number of craft that embarked on their missions in the opening years of the twenty-first century are still in the early stages of their exploratory programs. The European craft *Rosetta*, for example, still has to complete a ten-year journey before reaching its target, comet Churyumov-Gerasimenko. It will be in cruise mode for much of the journey. Then, on arrival in May 2014, it will move into orbit around the comet, release its lander, *Philae*, to the surface, and travel with the comet on its path around the Sun. It is the first mission to undertake a long-term exploration of a comet at close quarters.

Venus and Mars

Meanwhile, another European probe, *Venus Express*, will have arrived at Venus in 2006 and started its investigation. Its work will last for two rotations of Venus, about 500 Earth days. This mission was not even on the drawing board at the start of the century. It was born in 2001 when the European Space Agency put out a call to the scientific community for proposals that would reuse the design of the *Mars Express* to achieve a craft ready to launch in 2005.

Before *Rosetta* arrives in 2014, other craft will complete their work. *Mars Global Surveyor*, *Mars Odyssey*, *Mars Express*, *Mars Reconnaissance Orbiter*, and the two rovers *Spirit* and *Opportunity* all should have finished their studies of Mars; *Smart-1*'s job at the Moon and *SOHO*'s at the Sun will be complete; and *Ulysses* will have

◁ The two *STEREO* craft fly over Earth before commencing their studies of the Sun.

▧ *Rosetta* launches from Kourou, French Guiana, on board an Ariane 5 rocket.

The *Phoenix* lander craft, without legs, is upside down inside its aeroshell casing.

Once on Mars, *Phoenix*'s robotic arm will collect soil samples for on-site analysis.

made its final flyby over the Sun's poles. *Cassini* will have completed its four-year study of Saturn. Mercury will have been investigated for only the second time by space probe—the *Messenger* craft will make its three close flybys in 2008–09 and in early 2011 will go into orbit around Mercury.

A further mission, *New Horizons*, which set off for Pluto on January 19, 2006, has a particularly long journey ahead. Its launch speed was the fastest of any mission to date, and the craft passed by the Moon in an incredible nine hours to encounter Jupiter just one year later. It will not reach distant Pluto and its moon Charon until mid-2015; only then will its work begin. Its study of Pluto will be followed by up to ten years of travel and investigation among the objects of the Kuiper Belt.

New missions to Mars

The United States is to continue its study of Mars in the next few years with *Phoenix*, and *Mars Science Laboratory*. *Phoenix*, like its namesake the mythical bird, is being reborn. It is using equipment that was prepared for Mars but not sent, in particular the *Mars Surveyor 2001* lander, built in 2000, and two of its instruments. Its mission is to explore an arctic region of Mars, yet to be selected. The *Phoenix* launch is expected to be from Kennedy Space Center, Florida, in August 2007, with a scheduled arrival time of May 2008.

Human habitation

Phoenix will use aeroshell braking, a parachute, and thrusters to achieve a soft landing on the planet. The fixed lander will use its robotic arm to dig under the surface to access water ice, and then use its instruments to analyze samples. It is hoped that the probe will add to our knowledge of water on Mars and contribute to an assessment of the planet's potential for human habitation.

Mars Science Laboratory

Before the first decade of the twenty-first century is out, *Mars Science Laboratory* should arrive at Mars. It is a U.S. mission but with an international flavor because Russia, Spain, Canada, and Germany are all working on instruments for this rover. It will travel over the Martian landscape for at least a full Martian year (687 Earth days) and will be capable of moving at least 12 miles (20 km) from its landing site. Among other tasks it will collect soil and rock samples and analyze them for organic compounds that are essential to life as we know it.

Precision landing

The landing site will be selected in June 2009 from a list of 100 potential sites. The *Mars Reconnaissance Orbiter*, which then will be orbiting the planet, will look closely at these sites, and over the next two months the list will be narrowed down. The craft will then be launched in late 2009, to arrive in October 2010. *Mars Science Laboratory* will be the first planetary mission to use precision landing techniques; it will steer itself toward the surface and, when above its desired location, deploy its parachute. In the final minutes the mother craft will lower the rover to the surface on a tether.

Moon missions

A number of countries are planning to send missions to Earth's Moon. *Chandrayaan-1* is India's first venture into planetary space science; the country's first satellite was launched in 1975, and since then it has developed and launched many more, including a number of Earth observation and remote-sensing satellites. *Chandrayaan-1* is planned for launch as early as 2007 on board India's rocket, Polar Satellite Launch Vehicle. It will work in polar orbit for two years, helping to unravel the mysteries of the origin and evolution of the Moon and solar system. European instruments,

An impression of *Mars Science Laboratory* roving across the Martian surface.

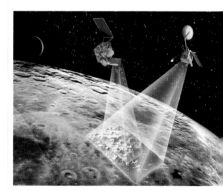

Chandrayaan-1 (*left*) maps lunar terrain alongside *Lunar Reconnaissance Orbiter*.

identical to those on the lunar craft *Smart-1*, will be on board.

Japanese lunar exploration

Japan is preparing two missions for the Moon, *SELENE* and *Lunar-A*. *SELENE* may be launched in 2007; it consists of three separate units: the main orbiter, and two small satellites, one for communication relay, the other for investigations of the Moon's position. The main craft, carrying thirteen instruments, will observe and investigate the Moon from its circular orbit for about one year. The launch of *Lunar-A* has been postponed a number of times, but this is not unusual in space exploration scheduling. It is anticipated that the craft eventually will be launched, reach its lunar orbit, and get to work imaging the Moon while its surface penetrators spend a year monitoring moonquakes. The two penetrators, which are missile-shaped cylinders 35 inches (90 cm) long, will be released individually to impact the Moon and burrow into its surface.

Lunar-A orbits the Moon after the release of the first of its two penetrators (left).

Chinese missions

One other orbiter will also be launched to the Moon in 2007. This is China's *Chang'e 1*, whose name comes from a Chinese legend about a young fairy that flies to the Moon. Its aim will be to test its technology for future missions as well as study the lunar environment and map its surface. Another possible future mission would collect soil samples and conduct tests in preparation for a manned Moon base.

Lunar Reconnaissance Orbiter

The United States aims to launch its *Lunar Reconnaissance Orbiter* to the Moon from the Kennedy Space Center in 2008. Its objective is to collect data needed for planning future missions to the Moon, such as sending men and women to the Moon, possibly as early as 2018, in preparation for an eventual manned mission to Mars. The United States is also looking into the

Chang'e 1 deploys its stereo camera during its one-year orbit of the Moon.

possibility of dropping landers on to the south polar region of the Moon, and returning samples to Earth.

BepiColombo

Europe's next planetary target is Mercury, and it is being joined by Japan in this venture. The craft is *BepiColombo*, which is named after the Italian scientist who, among other things, suggested a probe could get close to Mercury by using a gravity-assist flyby of Venus. *BepiColombo* consists of two separate craft. The first, *Mercury Planetary Orbiter*, is called a European craft; this will study the surface and internal composition of the planet. The second, called *Mercury Magnetospheric Orbiter*, is Japanese and this will study the planet's magnetosphere (the volume of space around the planet that is dominated by its magnetic field).

Toward the Sun

BepiColombo is not only Europe's first mission to Mercury but also the first time Europe has sent a craft to a hot region of the solar system; all of its previous craft have been to cold targets. The journey direction is also a first; for the first time a European craft must brake against the Sun's gravity, which increases with proximity, rather than accelerate away from it, as happens when a craft travels away from the Sun to the outer solar system. The probe's scientific instruments, which will work to throw light on the composition and history of Mercury,

Lunar Reconnaissance Orbiter directs its platform of instruments toward the Moon.

The *BepiColombo* craft arrive at Mercury; the European craft is in the foreground.

as well as the other inner planets, were selected in November 2004 and production started soon after.

BepiColombo's two craft should start their journey in 2012 aboard a Soyuz-Fregat rocket from Kazakhstan, although the launch schedule is not certain. They will reach Mercury more than four years later, after flying by the Moon and Venus, and will move into polar orbit, from where they will investigate the planet for one year.

⊡ A *STEREO* craft studies a huge eruption of gas from the Sun's outer atmosphere.

⊡ STEREO's stereoscopic images of solar eruptions will help to explain their origin.

STEREO

A two-year U.S. mission to the Sun called *STEREO* (Solar Terrestrial Relations Observatory) is scheduled to launch in summer 2006. It consists of two craft, one to be placed ahead of Earth in its orbit around the Sun, and one behind. Together they will image the Sun and the solar wind in three dimensions to improve our understanding of space weather and the Sun's impact on Earth.

Planet-C

Japan's first mission to Venus, *Planet-C* is an orbiter designed to study the planet's atmosphere, particularly the rotation of the upper atmosphere. It will also measure temperatures and look for lightning. Its launch is expected in early 2008 for an arrival at Venus in September of the following year.

Dawn

Some missions that have been proposed, or even built, do not make it to the launch pad. The U.S. *Dawn* mission was in the last few months of preparation for its May 2006 launch when the project was halted because of technical and managerial problems. The craft was to investigate the two largest asteroids, Ceres and Vesta, which are both found in the Asteroid Belt between the orbits of Mars and Jupiter. It was scheduled to arrive first at Vesta in July 2010 and then, after a year, move off to Ceres, to arrive in August 2014 and stay until July 2015.

JIMO

JIMO (Jupiter Icy Moons Orbiter) is another mission that has been put on hold. Its launch, once tentatively scheduled for 2015, is no longer on the U.S. launch program. It is possible, however, that this or a similarly designed craft will eventually travel to Jupiter. *JIMO* was to have been the first long-distance space mission to benefit from a newly-designed propulsion system, which allows probes to travel further and faster, as well as explore more efficiently. Extra power would have allowed *JIMO* to make a sustained study of three of Jupiter's largest moons, Europa, Ganymede, and Callisto, in turn.

Looking into space

A host of space telescopes, which will look into deep space as they orbit Earth, are presently under development for launch in the decade ahead. The infrared telescope Herschel, to be built in France, Germany, and Italy, will look at some of the coolest and most distant objects in the Universe, and so help further our understanding of how stars and galaxies are born. Darwin will survey 1,000 of the closest stars, searching for Earthlike planets, and Gaia will make a three-dimensional map of the Milky Way by surveying more than 1,000 million of its stars. XEUS (X-ray Evolving Universe Spectroscopy), which will consist of one spacecraft carrying a mirror, and another carrying detectors, is set to search for the first giant black holes.

The future for Mars

Mars is the focus of the long-term plans of a number of nations, namely the United States and a consortium of nine European countries working with Canada. By the start of the third decade of the twenty-first century, new planetary orbiters, rovers, and landers will be sent out from Earth. Airborne vehicles such as aircraft or balloons could also be used, and an astrobiology laboratory craft may conduct a robotic search for life.

Manned landings on Mars

Miniaturized science instruments on the surface, and deep-drilling systems to extend hundreds of yards beneath the surface, will have been employed, or be ready for work, and the first Martian samples will have been returned to Earth for analysis. Craft from the United States and Europe will increase in complexity over time, culminating, if all goes well, in at least one human expedition to Mars by 2030.

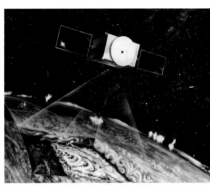

In this impression, *Planet-C* uses infrared sensors to study motion in Venus's clouds.

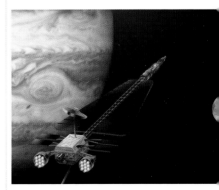

The *JIMO* craft is designed to use a small nuclear reactor to power electric thrusters.

Glossary

Aerobraking

Drag of atmosphere slows spacecraft

Aerobraking A maneuver whereby a spacecraft uses the atmosphere of a planet or some other body to slow down and change its orbit. For example, a spacecraft moving from a high circular orbit into a lower circular orbit will first use its onboard power to move onto an elliptical orbit. The lowest point on the elliptical orbit will take the craft inside the planet's atmosphere, where the craft is slowed by atmospheric drag. As a result, the high point of the orbit is lowered on this and successive passes through the atmosphere. Consequently, the low point of the orbit is raised out of the atmosphere and the desired lower circular orbit is achieved. The U.S. *Mars Reconnaissance Orbiter* used aerobraking after moving into orbit around Mars in March 2006.

Aeroshell A heat shield fitted to the outside of a spacecraft to protect it from the high temperatures experienced when aerobraking.

Antenna boom A rod-shaped device used on spacecraft for transmitting and receiving radio waves.

Atmospheric circulation The large-scale and global movements of the gases of a planet's atmosphere.

Aurora A colorful, glowing light display seen in the atmosphere around the polar regions of planets such as Earth, Jupiter, and Saturn. An aurora is produced when solar wind particles enter the upper atmosphere and interact with its gas, which then glows.

Axis of rotation The imaginary line that passes through the center of a body such as a planet or star, and about which that body rotates.

Capsule A small spacecraft, usually part of a bigger craft, used for carrying an astronaut or some other cargo.

Cosmic ray Highly energetic subatomic particles that speed through space at close to the speed of light.

Ecliptic plane The plane of the Earth's orbit around the Sun. The orbits of Earth's Moon and the planets, with the exception of Pluto, lie very close to the ecliptic plane.

Elliptical orbit An orbit is the path one body takes about another. On an elliptical orbit the path marks out the shape of an ellipse (a regular oval).

ESA European Space Agency.

Flyby The passage of a space probe past a planet or moon, where the craft neither orbits the body nor lands.

Gravitational field The region of space surrounding an object within which the object's gravitational pull is experienced by another object.

Gravity-assist A maneuver whereby a spacecraft flies past a planet and takes up a small fraction of the planet's orbital energy. The additional energy allows the craft to change direction and speed, or both. Due to its loss of energy, the planet slows imperceptibly and moves closer to the Sun. The gravity-assist maneuver reduces significantly the amount of fuel a craft needs to carry. It has been used on many space missions including that of Cassini-Huygens to Saturn (*see below*). After launch, Cassini-Huygens made two gravity-assist passes of Venus and one of Earth, which together propelled it to the outer solar system. A final gravity-assist flyby of Jupiter boosted it on to its destination, Saturn.

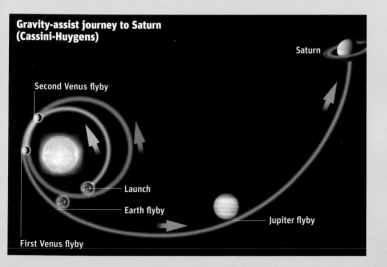

Gravity-assist journey to Saturn (Cassini-Huygens)

Saturn

Second Venus flyby

Launch

Earth flyby

Jupiter flyby

First Venus flyby

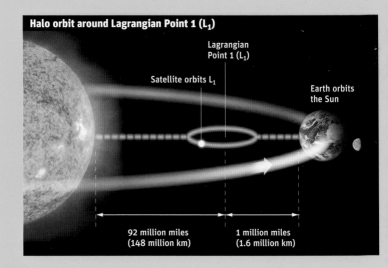

Halo orbit around Lagrangian Point 1 (L₁)

Lagrangian Point 1 (L₁)

Satellite orbits L₁

Earth orbits the Sun

92 million miles (148 million km)

1 million miles (1.6 million km)

Halo orbit A spacecraft can follow an orbit around Lagrangian point L₁ or L₂. (*see* Lagrangian points). Such an orbit, which is around an empty point in space, is called a halo orbit. The *SOHO* spacecraft observes the Sun from its halo orbit round L₁. The Herschel craft, once launched, will move into its halo orbit around point L₂. Craft following halo orbits fire their thruster rockets regularly to maintain their orbital path.

Heliopause The outer boundary of the heliosphere, which is the vast volume of space surrounding the Sun and in which the solar wind has an effect.

Interplanetary dust Dust particles within the solar system, which in the main were released by comets as they move close to the Sun.

JAXA Japanese Aerospace Exploration Agency.

Lagrangian points The five locations in space where, despite the gravitational pull of two more massive bodies, a small body such as an asteroid or spacecraft can maintain a stable orbit. The five points are named L₁ through to L₅. In the Earth–Sun system, (*see artwork opposite*) a spacecraft placed at these points remains stationary relative to the two massive objects. However, any disturbance to a craft at points L₁, L₂, and L₃ will cause it to move away, unless it powers itself back. A craft at L₁ gets an uninterrupted view of the Sun, one at L₂ looks into deep space. L₃, on the opposite side of the Sun, is unused. L₄ and L₅ are potential locations for future habited spacecraft.

Magnetic field The region surrounding a magnet, or a body with magnetic properties such as Earth, where its magnetic force is felt.

Magnetometer An instrument used for measuring the strength and direction of a magnetic field.

Magnetosphere The volume of space around a planet where solar wind particles are controlled by the planet's magnetic field rather than the Sun's.

NASA National Aeronautics and Space Administration.

Polar orbit The path followed by a spacecraft passing over, or close to, the poles of a planet or the Sun.

Radar mapping The practice of bouncing radio waves off a planet or moon to map its surface. Bright areas on a radar image denote rough terrain, dark areas indicate smooth terrain.

Radiation belt A doughnut-shaped ring around a planet where particles are trapped by its magnetic field.

Solar activity cycle The Sun's 11-year cycle of activity, notably the production of sunspots and flares. The number of sunspots rises and falls over the cycle. About 120 sunspots per month occur at solar maximum, and about six at solar minimum.

Solar array A panel made up of solar cells, which convert the solar energy that falls on them into electrical energy. Solar arrays are used to power probes required to travel no further from the Sun than Mars.

Solar atmosphere The gases that lie beyond the visible surface of the Sun. Immediately above the surface is the chromosphere; beyond this is the extensive corona.

Solar flare An explosive release of energy just above the Sun's surface.

Solar orbit The path taken by any body traveling round the Sun, such as Earth and the *Ulysses* space probe.

Solar wind The flow of fast-moving energetic particles that escape from the Sun.

Spectrometer An instrument that splits radiation, such as light rays, infrared, and x-rays, into a spectrum, and then identifies and measures the properties of that radiation (and therefore the object it came from, such as a star or planet).

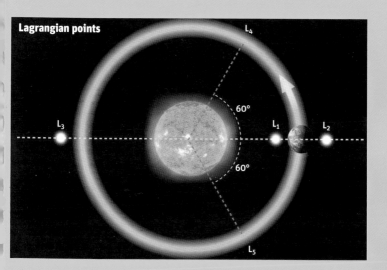

Lagrangian points
L_4
L_3
$60°$
L_1
L_2
$60°$
L_5

Index

Picture Credits

EXPLORING THE SOLAR SYSTEM

(t=top, b=bottom, l=left, c=center, r=right, fr=far right)

China National Space Administration (CNSA): 86b

European Space Agency (ESA): Front cover, 9, 10t, 11, 16, 18b, 32, 39t, 39b, 45, 46, 47, 54t, 70t+b, 74t, 83, 87b

Getty Images: 78t NASA/Getty Images

Indian Space Research Organisation (ISRO): 85b

Japanese Aerospace Exploration Agency (JAXA): 10b, 80b, 86t, 89t

National Aeronautic Space Administration (NASA): 2, 12t, 14t, 15, 20, 21t+b, 22, 23t+b, 24t+b, 25, 26b, 27, 31t, 33, 34, 36t, 37b, 38, 41, 43t+b, 48, 50b, 51, 52, 55, 56t+b, 58t, 59t+b, 62t+b, 63, 71b, 73, 77b, 80t, 81b, 82, 84t+b, 85t, 87t, 88t+b, 89b

RIA Novosti: 17, 19t+b, 28, 29, 35b, 49

Science Photo Library: 6 NASA/SPL; 12b NRSC Ltd/SPL; 13t Space Imaging/SPL; 13b Space Imaging/SPL; 14b NASA/SPL; 16 Christian Darkin/SPL; 18t BMDO/NRL/LLNL/SPL; 26t NASA/SPL; 30 BMDO/SPL; 31b Christian Darkin/SPL; 35t RIA Novosti/SPL; 36b NASA/SPL; 37t NASA/SPL; 40 Ton Kinsbergen/ESA/SPL; 42 NASA/SPL; 44t NASA/SPL; 44b JISAS/ Lockheed/SPL; 50t NASA/SPL; 53t NASA/ SPL; 53b NASA/SPL; 54b ESA/SPL; 57 NASA/ SPL; 58b NASA/SPL; 60 David Ducros/SPL; 61 NASA/SPL; 64 Peter Ryan/ SPL; 65t NASA/SPL; 65b US Geological Survey/ NASA/SPL; 66 NASA/ SPL; 67t Julian Baum/SPL; 67b NASA/SPL; 68t NASA/SPL; 68b NASA/SPL; 69 NASA/ SPL; 71t ESA/ SPL; 72 NASA/SPL; 74b ESA/SPL; 75 Novosti/SPL; 76t NASA/SPL; 76b NASA/ SPL; 77t NASA/JPL-Caltech/ UMD/SPL; 78b Erik Viktor/SPL; 79t John Hopkins University Applied Physics Laboratory/ SPL; 79b NASA/SPL; 81t NASA/KSC/SPL; Back cover Jerry Lodriguss/SPL

BOX AND COMPONENTS

Box front: Detlev Van Ravenswaay/SPL

Box back: Detlev Van Ravenswaay/SPL

Box liner: Jerry Lodriguss/SPL

Wall chart: Detlev Van Ravenswaay/SPL; Victor Habbick Visions/SPL

Mission markers: top row, l NASA; c ESA; r ESA; second row, l JAXA; c ESA; r NASA; third row, l NASA; c NASA; r NASA; fr NASA; fourth row, l NASA; c NASA; r NASA; fr NASA; bottom row, l NASA; c NASA; r NASA; fr NASA

Mobile planets: Victor Habbick Visions/SPL